GRAPHENE

GRAPHENE
The Route to Commercialisation

James Baker
James Tallentire

JENNY STANFORD
PUBLISHING

Published by

Jenny Stanford Publishing Pte. Ltd.
Level 34, Centennial Tower
3 Temasek Avenue
Singapore 039190

Email: editorial@jennystanford.com
Web: www.jennystanford.com

British Library Cataloguing-in-Publication Data
A catalogue record for this book is available from the British Library.

Graphene: The Route to Commercialisation

Copyright © 2022 by Jenny Stanford Publishing Pte. Ltd.

All rights reserved. This book, or parts thereof, may not be reproduced in any form or by any means, electronic or mechanical, including photocopying, recording or any information storage and retrieval system now known or to be invented, without written permission from the publisher.

For photocopying of material in this volume, please pay a copying fee through the Copyright Clearance Center, Inc., 222 Rosewood Drive, Danvers, MA 01923, USA. In this case permission to photocopy is not required from the publisher.

ISBN 978-981-4877-87-9 (Hardcover)
ISBN 978-1-003-20027-7 (eBook)

Contents

Preface		ix
Introduction		1
1. The Innovation Revolution: From Closed to Open Innovation		**13**
1.1	Early Experience in Open Innovation	16
1.2	The Wildcat Is Let Loose	17
1.3	The Oxford Connection	19
1.4	Bringing the Wildcat Experience to Manchester	20
1.5	Working with SMEs	20
2. Graphene: The New Kid on the Block		**25**
2.1	The Accidental Pioneers	26
2.2	Playfulness Rewarded with a Nobel Prize	28
2.3	Graphene Becomes a Global Icon	29
3. Graphene's Great Expectations		**33**
3.1	"Plastic Planes": They Will Never Take Off!	34
3.2	World's First Graphene Plane	36
3.3	Richard Branson Backs Graphene	38
3.4	Graphene Drives Auto Innovation	39
3.5	Better Batteries, Supercapacitors and Hybrids	40
3.6	World's First "Graphene-Bodied" Car	40
3.7	The Covid-19 Catalyst	41
3.8	Fast Lane and the Slow Lane	41
3.9	In the Fast Lane: The Rubber Experience	42
3.10	Keep on Running: The World's First Graphene Shoes	43
3.11	Need for a New Model Approach	46
4. Overcoming the Challenge: A New Model Approach		**49**
4.1	The Innovation Journey: The Long and Winding Road	50
4.2	Getting Past the Hype	51
4.3	Electronics Possibilities	52

4.4	Technology Readiness versus System Readiness	52
4.5	Creating the Manchester Model of Innovation	54

5. Building the First Home for Graphene: the NGI — 59

6. Steal with Pride: Creating the GEIC — 67
6.1	The Missing Piece	67
6.2	Football Opens Up Leftfield Funding Opportunities	68
6.3	More Industrial than Academic	69

7. Graphene City and Beyond: Building an Innovation Ecosystem — 75
7.1	Recognising the Size of the Prize	75
7.2	Developing a City Centre Innovation Hub	77
7.3	Creating an Out-of-Town Manufacture Hub	79
7.4	Made in Manchester	80
7.5	Winning Hearts and Minds	81
7.6	Levelling up Graphene Opportunities	82
7.7	Materials for "Sustainable Manufacture"	82
7.8	Regional Leadership in 2D Materials	84

8. Making It Happen: Industry Engagement — 87
8.1	From an Academic-Led to Industry-Led Culture	87
8.2	A Full House	89
8.3	Bridging the Gap—Supporting Start-Up and SMEs	94
8.4	Setting a "Gold Standard" for Graphene	95

9. Getting Graphene Ready: Adopting the Manchester Model of Innovation — 99
9.1	Meet the Graphenes	99
9.2	Push or Pull	103
9.3	Plugging the Graphene Skills Gap	104
9.4	PhDs in Graphene	104
9.5	Innovation against the Clock	105
9.6	Enterprise Case Study: Hacking a Path to Innovation	106
9.7	Innovation and Skills	107
9.8	Enterprise Award Helps Kick-Start Innovation	108

10. Creating an Icon: The Making of a Global Brand — 111

- 10.1 The One That Got Away — 111
- 10.2 Birth of "The Baby" — 112
- 10.3 Losing Out to the USA — 113
- 10.4 Reviving a Neglected Heritage — 114
- 10.5 A Manufacturing Retreat — 116
- 10.6 The Secret to Graphene's Universal Appeal — 117
- 10.7 The Quest for a Story with Universal Appeal — 117
- 10.8 The Game-Changer—Winning the Nobel Prize — 118
- 10.9 PR: Vital Ingredient in the Commercialisation Mix — 120
- 10.10 Managing the Story, Creating a Unique Vision — 121
- 10.11 Sharing Graphene's Alchemy: The Birth of the Beacons — 122
- 10.12 Graphene, "An Icon of UK Innovation": Building a Global Reputation — 127
- 10.13 Flying the Flag for UK Innovation — 129
- 10.14 Graphene in China and India — 130
  - 10.14.1 Graphene Goes to the USA — 132
  - 10.14.2 Graphene's Gravitas — 133

11. The New "Gold Rush": Graphene's Research Renaissance — 135

- 11.1 Living in the Graphene Age — 135
- 11.2 The Graphene Gold Rush — 136
- 11.3 Graphene's Not Dead — 137
- 11.4 Let's Do the Twist — 138
- 11.5 On the Wings of a Butterfly — 140
- 11.6 Graphene v the Graphenes — 141
  - 11.6.1 Making the Intelligent Club Sandwich — 143
- 11.7 Once Lost, Now Found: Magnetic 2D Materials — 143
- 11.8 Rising Star in the Graphenes Family: Graphene Oxide — 144
- 11.9 Going for a Spin — 145
- 11.10 The Speed of Light — 147
- 11.11 Evolution, Not Revolution — 148
- 11.12 Graphene Goes Green: Responding to Climate Change — 149
- 11.13 Reinventing the Battery — 149

11.14	Building the Future: Graphene in Construction	151
11.15	Responding to Covid-19	153

12. Future Histories: Graphene Innovation after Covid-19 — 163
- 12.1 Demand-Led Approach — 164
- 12.2 At the Tipping Point — 166
- 12.3 Future Histories — 167

Index — 169

Preface

The common perception regarding British innovation is that the UK is great at invention but we always underperform when we try to commercialise these ideas.

This opinion has been articulated by a number of commentators. For example, in a survey conducted by *The Engineer* magazine (May 2018), it was revealed that "sentiment regarding the UK's ability to commercialise technology was overwhelmingly negative, with nearly 85 per cent feeling the country is falling short in some respect". However, Professor Tom Nicholas, from Harvard Business School, in a contribution to the *Cambridge Economic History of Modern Britain* (2014)[1] says: "Britain has not failed when considering the function of invention. Rather, Britain has a long tradition of excellence in invention and in basic science".

This positive statement is then followed by a litany of great British discoveries and inventions, including the gyro-compass, the jet engine, the float glass process (the most common manufacturing process of flat glass sheets), cotton textile machinery (that helped trigger the original industrial revolution in Manchester and Lancashire), DNA, the CT Scanner, and the World Wide Web. I am sure the readers of this book could add many, many more equally seminal innovations—for example, the first modern computer was built in the UK (Manchester in fact) and the following generation of computers were created to serve Britain's nuclear energy scientists, another field where we acted as post-war global pioneers. But the point Professor Nicholas goes on to make is that despite these areas in which Britain has displayed such competence and distinctiveness, "the management of technology has typically been weak" and the UK has long suffered structural flaws in its commercialisation environment.

---

[1]Taken from T. Nicholas. Technology, Innovation and Economic Growth in Britain Since 1870 [Prepared for the Cambridge Economic History of Modern Britain, 2014], in: *The Cambridge Economic History of Modern Britain: Volume II. 1870 to the Present* (R. Floud, J. Humphries, P. Johnson, eds.), Cambridge University Press, pp. 181–240.

There is a web of interconnected reasons for this science and innovation deficit, but one we believe is vitally important is the need to have the right 'ecosystem to support R&D' as a way to harness inventions developed domestically. This book looks at how Britain—and indeed Greater Manchester, the region once known as 'Cottonopolis' for its revolutionary textile technologies that made global impact—has not only given the world another great science breakthrough in the form of graphene, the world's first 2D material, but also looked to retain its commercialisation in the place of discovery.

This ambition is being achieved by developing a unique translational environment that has its base in the very institution where graphene was first isolated, The University of Manchester. However, a campus-based R&D accelerator is very limited and so the Manchester graphene community has been joined by civic organisations, plus a diverse range of business partners at both national and international levels. To add to this mix is the UK government, who were early champions of graphene and hoped the new Manchester model of innovation would break the old cycle so that 'Discovered in Britain' would this time also lead to a boast of 'Made in Britain'.

To help tell the complex and fascinating story behind the commercialisation of graphene at Manchester, we have been fortunate to have access to key players in this narrative and who have generously given their time and first-hand testimonials. We are very grateful to these pioneers, innovators, senior leaders, and key stakeholders who have candidly explained their role in supporting graphene on its journey from scientific breakthrough to the real-world application. They have revealed just how challenging this bold quest has been with a whole series of political, financial, reputational and commercial considerations having to be negotiated all along the way. These contributors—and many, many more colleagues and partners whom we were not able to highlight or namecheck—have acted as architects to create this bespoke ecosystem, which has been dubbed 'Graphene City'. It is still very early days for this proposed and ambitious 'innovation utopia', which has been inspired by Nobel prize-winning science and the concept of open innovation, but a firm foundation has now been laid—and anyone who is a part of this game-changing project becomes dedicated to its delivery.

What has also become clear from studying Manchester's graphene story is the fact that this invisible 2D material has an unexpected ability to be democratic and embracive. People from all walks of life and different backgrounds have been fascinated by this 'wonder material' and want to know more about its extraordinary potential. We believe this allure all begins with its origin story that tells of two brilliant and curious scientists—Andre Geim and Kostya Novoselov—who experimented on a Friday night using nothing more than sticky tape to isolate a brand new material. And the intelligent story-telling and brand building that then followed cannot be underestimated because it has made a pivotal contribution to driving graphene along the route to commercialisation by attracting the right actors at the right time to keep momentum going.

Finally, a huge acknowledgement and big thanks must go to our immediate families, whose patience has made the writing of this book possible. Countless lost weekends have been tolerated and thankfully forgiven.

James Baker and James Tallentire
November 2021

Introduction

By James Baker and James Tallentire

James Baker, Manchester: Throughout my career, I have been driven by the belief and importance of "making a difference" in what I do and in any of my outputs and outcomes that I am looking to achieve. This book is dedicated to graphene, which I first came across in 2013 whilst working in a large corporate organisation. Graphene is an example of a new material which truly has the potential to make a real difference not only to the life in which we all live in today but also to create significant value for the UK. Working in a team has always been important to myself and has significance in the way that I operate—and this book is really about the people within the various stakeholder organisations—both academic and commercial—who have the opportunity to bring graphene from the lab to the marketplace. Not only is this book about graphene as a new material, but also it looks at new business and innovation models, and a more collegiate way of working.

This book explores how we might better bring together the best of academia and industry to take some of the great inventions that take place in the UK and how we can bring them to the marketplace. The UK has a history of great discovery but has not always been in first place when it comes to creating the supply chain and achieving value in terms of new products, jobs and factories. It has also been a challenge that new materials have typically taken many, if not tens of

---

Graphene: The Route to Commercialisation
James Baker and James Tallentire
Copyright © 2022 Jenny Stanford Publishing Pte. Ltd.
ISBN 978-981-4877-87-9 (Hardcover), 978-1-003-20027-7 (eBook)
www.jennystanford.com

years to go from discovery to the first products in the marketplace. Graphene, at only 17 years "young" since first discovery in 2004, is still in its teenage years and therefore is still in the early phases of commercialisation. This book is dedicated to tell the story of the last few years where graphene truly has started to increase its pace towards commercialisation and is now approaching a "tipping point" of products and applications into the mainstream.

Graphene was first isolated at The University of Manchester in 2004. I was first introduced to graphene while working for a large FTSE 100 aerospace and defence company. My background is from industry, where I have worked for over 25 years, generally involved in the commercialisation and the pulling through of technology. In my final years in industry, I became involved in an exercise looking at the development of "futures"—business and scenario planning looking forward to developments in 20 years' time. We ran a series of events that brought together the leaders of the organisation to try and understand not only what the future might look like but also how we might react to future demands and challenges.

It was during this very exciting activity looking at the future, which was great as generally you cannot really get it wrong because, frankly, nobody knows what the future might hold—and it was then that I first came across the existence of this wonderful material called graphene. At that moment, I had very little knowledge about graphene but we used it as an example of how a new material might be applied in the future. A little background reading about graphene revealed the superlative properties associated with this original and first 2D material, like the fact it is 200 times stronger than steel, more conductive than copper, strong, flexible and transparent. All these properties could make a real difference to many different products and applications in many different market sectors.

What really attracted me to graphene was its "multifunctional" properties—the ability to have, for example, both strength and also conductivity (thermal and/or electrical). This material really had the opportunity to "make a difference" and be truly "disruptive" not only in the markets I was then focused on but also in many other markets and sectors beyond aerospace and defence. Historically it could be argued that many new materials were often pulled through by the defence market which as early adopters were prepared to take the risk (and the benefits) that a new material might provide.

Interestingly, we have seen a change over recent years in the pull through of new technologies; typically, they are now driven much more quickly and effectively by the commercial sector. Graphene is an example of a material that would have caused great interest for defence and aerospace communities like carbon fibre before it. However, if you look historically at examples like carbon fibre it was probably 25 years from discovery through to the first products appearing and applications in the market. These products were primarily from aerospace and defence applications but also from sporting goods like tennis rackets and Formula One applications. With the challenges that we might face in the future, we realised that we would need a different way of working if we were to successfully pull through technologies for aerospace and defence in the future.

It was then by coincidence that I received a call, initially via a social media post from The University of Manchester regarding a potential job opportunity, to work with this so-called 'wonder material' graphene. I followed up with a conversation with Professor Colin Bailey CBE, at that time Dean of the Faculty of Science and Engineering at The University of Manchester—and now President and Principal of Queen Mary, University of London—whose enthusiasm for graphene's potential further inspired me. From this engagement I was hooked; so I decided to leave my industry-based career, which I had undertaken for over 25 years, to become directly employed in the higher education sector as an employee of The University of Manchester—and the 'home of graphene'—working alongside business development director Ivan Buckley, who was also from industry and has now been engaged with graphene commercialisation for over 10 years. It seemed a chance to truly make a difference and a great opportunity to leverage my previous skills and experience and to bring this wonder material graphene into potential products and applications for the future.

The Tipping Point

During my time at The University of Manchester, I have presented the story of graphene and the work that the University is doing to commercialise this family of new 2D materials. At many conferences and events, I have presented on: "Graphene—The route to commercialisation"; then in 2018 I was approached by a publishing company who saw great interest and value in telling the story of graphene commercialisation and so I decided to co-write this book. It seemed to me an excellent opportunity to explain what drove me personally and, also as a community, what we have done to help bring graphene towards the "tipping point" as we approach the end of 2021.

During my career—which has included leading a R&D community at the British multinational defence and aerospace company BAE Systems—I have been lucky enough to witness numerous innovations approaching their own respective tipping points, for example in autonomous vehicle systems which was a technology that was moving along very slowly and with few supporters and budgets available, to progress to what we are seeing currently with for example the Tesla cars on the roads today. Such tortoise-like progress can be accelerated by an event or events until a critical point is suddenly reached and it seems everybody is talking and wanting to be involved in such a disruptive system. In the early days of autonomous systems, we had many cynics who said there were no requirements for such systems. In aerospace, our ideas were, in fact, received with scepticism to suggest that we might fly an aircraft completely without a pilot—and yet today, of course, unmanned aerial vehicles (UAVs) often called drones are quite commonplace in both military and civic applications. I believe the best way to accelerate a new technology into market is to adopt an open and sharing approach to innovation with industry and academia working as symbiotic R&D community. As far back as 2011, I was publicly advocating that UK Plc needed to move away from a culture of splendid isolation and do things differently if it were to remain competitive in a fast-changing world with ever shifting geopolitics and technological advances. In a blog in the Huffington Post [1], I wrote:

"Open innovation therefore is not just a business model. It's a mindset that recognises that Britain cannot remain at the cutting edge of innovation without collaboration across industry and academia. There is a wealth of ideas in this country and to make the most of them we need to be mature in our approach and thinking and recognise that, even for a company the size of BAE Systems, it is simply no longer possible or desirable for a single company to work on them in isolation."

When at BAE Systems we undertook the "futures" activity, we saw many similar case studies when people had great technologies or ideas but had to overcome the many barriers or obstacles during the journey from ideas through to concepts through to commercialisation and success. The need to break down these hurdles has always been a key driver in my thoughts, actions and operations. To push an innovation in its early years and overcome obstacles by, for example, developing standards and also carefully managing communications and engagement to overcoming any hype or industry scepticism. This book is dedicated to not only the team at The University of Manchester but also other teams around the world who have overcome the significant obstacles to get graphene to the point where it is today and close to achieving the tipping point of commercialisation. When we look back in history, we see a number of these moments of crossover and all the challenges and inertia that has to be overcome. Henry Ford once famously said if you were to ask people of his day what innovation they would most want they would reply 'a faster horse'—instead Ford gave them what they never knew could exist, an affordable automobile.

So, it is with great excitement that I co-write this book and I believe my 25 years in industry provides a great background knowledge, i.e. taking new technology and bringing that innovation through into products and applications. It is seven years since I joined The University of Manchester working on the graphene project and there is no better example of a new material that is probably going to be the quickest new material to market—but not only that, a material that truly has the potential to disrupt many sectors and areas and make the difference and start to rewrite a book of "future histories".

In deciding to co-write this book, I joined up with my colleague James Tallentire, who has helped bring the graphene narrative to life through his significant skills and background in communications and marketing. Together I believe we can write and produce a narrative that not only describes the excitement and the ability of graphene to truly make a difference but also to describe the excellent activity not just in graphene but the whole advance materials sector across The University of Manchester. The team is working not only on the science but in developing the commercialisation in the National Graphene Institute (NGI) and the Graphene Engineering Innovation Centre (GEIC) at Manchester.

By bringing the ecosystem together, we truly can deliver on the "Graphene city" vision.

***

James Tallentire, Manchester: I have worked in the higher education sector for the past two decades supporting leadership teams at Staffordshire University, Birmingham City University and The University of Manchester with their corporate communications and engagement. Although these are very different universities—especially as Manchester is a research-led institution—all have shared a similar commitment to support the economic and social development of their respective city-regions through applied research and innovation, as well as creating the talent and leaders needed to make positive change happen. Most recently I have been an advocate of the lab-to-market innovation being pioneered by The University of Manchester's world-class advanced materials portfolio, including an ambitious ecosystem that brings academic, civic and business leaders together to commercialise graphene. What we now proudly describe as the Manchester Model of Innovation.

This work to be an economic catalyst, I believe, is an increasingly vital contribution being made by the higher education sector. Indeed, Nunes, Tomé and Duarte Pinheiro (2013) [2] note that while many scholars agree that universities are generally engaged in delivering three fundamental missions—education, research and social endowment—others' researchers, such as Goddard (2009) [3], have also argued that a 'fourth mission' has emerged, the **economic mission**. This fourth mission is focused on stimulating the dynamics of regional and urban development. Boffo and Cocorullo (2019) [4] have also identified this fourth mission, in particular university spin-out enterprise which they describe as "… instruments designed to respond to social pressures towards accountability and establish a dialogue with the economy through the sharing of academic research findings." This ethos about the sharing of research closely echoes James Baker's discussion on open innovation.

In my career in strategic communications I have been involved in developing messaging and articulating clear propositions around the 'fourth mission' at Staffordshire University, Birmingham City University and The University of Manchester. During my time with

Staffordshire University I was very involved in promoting that institution's pioneering commitment to enterprise, which included three business villages at each of its then three main campus sites, based at Stoke-on-Trent, Stafford and Lichfield. Enterprises based in these incubator communities, many of which were set up by Staffordshire graduates, ranged from multi-media companies and filmmakers to electronics spinouts and even financial advisers. I personally worked with a number of the media and web companies as we looked to augment the university's own in-house communications function by commissioning the services of homegrown graduate talent in filmmaking and digital expertise.

To attract and nurture even more high-tech and innovative businesses to the city-region an ambitious step change came in the shape of the £282 million University Quarter (UniQ) initiative, which was led by the university in partnership with the local city council and Stoke-on-Trent's two FE colleges. The three education institutions were all situated within close proximity and therefore provided a focus for the UniQ project which aimed to rejuvenate an inner-city area of Stoke-on-Trent by creating an integrated learning community with a remit to drive up regional aspirations and skills to then create a talent supply to feed new high-tech starts-ups or innovative businesses. As the communications lead for the UniQ partnership I facilitated the engagement with internal stakeholders within the partnership—but the critical breakthrough came after persuading the then university secretary Bill Rammell [and later the Vice-Chancellor of the University of Bedfordshire] to visit the proposed UniQ site. This excursion paid dividends, because soon after Mr Rammell's fact-finding tour the project quickly received regional development agency funding to kick-start work on the University Quarter. From this starter funding the project soon gained momentum and has since developed into an award-winning community campus that has significantly contributed to renewal in this inner-city area.

At Birmingham City University (BCU), I was part of the award-winning Birmingham Made Me campaign to support the University's corporate ambitions to build a reputation in "innovation, design, invention and manufacturing" that was strongly anchored to 'Place'. The West Midlands is associated with a diverse range of makers that are each associated with iconic products, such as Jaguar Land Rover, AGA Rangemaster Group, Morgan Motor Company, Pashley Cycles, ACME Whistles, Emma Bridgewater, JCB, Dunlop, Triumph, Vax electrical goods-maker, and Brooks England, the bicycle saddle-makers [5]. These businesses would all become associated with the Birmingham Made Me project which was the brainchild of West Midlands' businesswoman and politician Beverley Nielsen who recognised that successful brands were built on the symbiotic relationship with manufacturing innovation and good design. "Manufacturing is often a very visual business and you have to work with all sorts of different creatives," explained Beverley. Through her leadership role at Birmingham City University—Beverley is an Associate Professor and Director at the IDEA Institute, which works with local start-ups to bring them to market—I became the communications lead for the project's original launch campaign, an exciting role which included stakeholder engagement, events promotion and media relations. The Birmingham Made Me brand would go on to to include a jobs fair, entrepreneur pop-up shops and high profile expos that showcased the best in the region's design-led manufacture. In the initial phases of the project the priority was to galvanise support among key stakeholders in the city-region, including civic leaders, such as Sir Albert Bore, then leader of Birmingham City Council, and Jerry Blackett, CEO of Birmingham Chamber of Commerce, as well as building government-level contacts with contacts at the Department of Business Innovation and Skills [now the department for Business, Energy and Industry Strategy]. National and local decision-makers were linked in with the region's business and academic leaders through a mix of activities, including sponsored expos and presenting a policy 'white paper' on the need for government to invest in the UK's design led-manufacture as part of the 'real economy' entitled *Looking for Growth: Sack the Economists Hire A Designer*' [6] at a fringe event at the 2012 Conservative Party Conference. This work was supported by the advocacy of influencers, such as economist Vicky Pryce, the former Joint Head of the United Kingdom's Government Economic Service. Through the management of this network it then became possible to articulate how design-led manufacture was crucial to the success of the West Midlands economy, including the vital role of knowledge partners such as Birmingham City University who could provide creative innovation and a supply of talented and relevantly skilled graduates. The Birmingham Made Me proposition was also successfully pitched to regional and national media. For example, we successfully placed Beverly Nielsen and other BCU innovation experts on high profile news slots on BBC Radio 2, BBC Radio 4 and the BBC news flagship Newsnight.

I joined The University of Manchester in 2013 in a communications and marketing role which led to supporting the University's proposition around advanced materials as part of the corporate beacons campaign (see Chapter 10). This in turn led me to meet James Baker in his role heading the team that was dedicated to the commercialisation of graphene and 2D materials. Graphene's commercialisation has not only become part of the University's "economic mission" but is also shared with the city-region. As a consequence, the advanced materials communications function has gone way beyond the confines of our campus to include working closely with regional civic partners like MIDAS, Greater Manchester's own inward investment agency, and the office of the elected mayor Andy Burnham and the Greater Manchester Combined Authority, which has placed graphene into its Local Industry Strategy (see Chapter 7). Graphene is also championed at a national level by the UK government, in particular the Department of International Trade (DTI), which has profiled graphene as the latest icon of innovation in Britain's long history of applied science and invention.

This shared advocacy of graphene's scientific and economic potential means the University gets a bigger bang for its buck in terms of promoting its most iconic of research assets. This is an enviable position to be in; but the challenge has been developing a single voice and much work in recent years has focused on developing a coherent lexicon of messaging and targeted communications to ensure that Manchester is recognised as the "home of graphene" (see Chapter 10). The critical point I think would be relevant to the readers of this book is how careful and strategic management of Manchester's graphene story has provided powerful leverage to help innovation leaders to drive forward the commercialisation ambitions for graphene. As Alan Ferns, former Associate Vice-President, says in Chapter 10:

"Building profile with professional communications has opened up some funding opportunities and partnership opportunities that just wouldn't have come. I don't think that the politicians would have been interested if we hadn't been telling the story in the right way, I don't think some of the investors would have been interested unless it was something they wanted to attach their name to. I think [the communications around graphene] have played a big role."

Another interesting influence on the commitment to make graphene a commercial success was a story I first heard from Alan Ferns, who had worked hard to reconnect The University of Manchester to its almost forgotten heritage inventing another great innovation—the first modern computer by Tom Kilburn and his team of electrical engineers who were helped by pioneers like Alan Turing. This innovation was translated into a regionally-based computer industry—but over the decades this UK sector was to be eclipsed by computer innovation led in America and Japan. Can a lesson be learnt as we reach the tipping point in graphene's commercialisation? I believe this book provides the answer.

References

1. J. Baker To remain competitive we need to innovate a new way of doing business, *Huffington Post*, 08/11/2020. https://www.huffingtonpost.co.uk/james-baker/to-remain-competitive-we-_h_953565.html.
2. D. Marques Nunes, A. Tomé, M. Pinheiro. Regenerative universities? The role of universities in urban regeneration, 2013, DOI: 10.13140/RG.2.1.3298.0960. https://www.researchgate.net/publication/281591734_Regenerative_universities_The_role_of_universities_in_urban_regeneration.
3. J. Goddard. Reinventing the civic university. *NESTA Provocation*, 12, 9/2009. London: National Endowment for Science, Technology and the Arts. https://www.nesta.org.uk/report/re-inventing-the-civic-university/.
4. S. Boffo, A. Cocorullo. University fourth mission, spin-offs and academic entrepreneurship: Connecting public policies with new missions and management issues of universities, 2019 16, 125–142. https://www.researchgate.net/publication/332154700_University_Fourth_Mission_Spin-offs_and_Academic_Entrepreneurship_Connecting_

Public_Policies_with_New_Missions_and_Management_Issues_of_Universities.

5. Birmingham Made Me 2014: Manufacturing happens by design, not by accident. Birmingham Live, November, 2014. https://www.business-live.co.uk/economic-development/birmingham-made-2014-manufacturing-happens-8056397.B. Nielsen. Looking for growth: Sack the economists hire a designer, industrial policy white paper, 2012. https://bcuassets.blob.core.windows.net/docs/Idea_Bham_LFG_LowRes_Spread.pdf.

Chapter 1

The Innovation Revolution: From Closed to Open Innovation

By James Baker

Scientific research into graphene—the world's first 2D material—has become a global phenomenon, following its isolation in 2004 and then Nobel Prize recognition in 2010. If commercialisation of this breakthrough material was to be delivered in a way and at a pace that would meet the ever-growing economic and societal expectations being placed on this so-called 'wonder material', then the innovation model required would have to be special. In 2014, I found myself at the heart of an amazing team of people based at The University of Manchester who were determined to make graphene a commercialisation success story. I was able to draw deep on my own experience from industry and decided to advocate a very collaborative approach and to adopt "open innovation" as Manchester's own model for taking graphene from lab to market.

From early in my career, I came across the principle of "open innovation," which was first coined by Professor Henry Chesbrough of the University of California at Berkeley. In his seminal 2003 book *Open Innovation: The New Imperative for Creating and Profiting from Technology* [1], Chesbrough argued that there had been a paradigm shift from what he called Closed Innovation, which is a system that believes successful innovation requires control. So an organisation must generate their own ideas, develop them, make them, market

---

Graphene: The Route to Commercialisation
James Baker and James Tallentire
Copyright © 2022 Jenny Stanford Publishing Pte. Ltd.
ISBN 978-981-4877-87-9 (Hardcover), 978-1-003-20027-7 (eBook)
www.jennystanford.com

them, and so on. The new paradigm was the adoption of Open Innovation, which assumes businesses make the best use of both internal and external ideas it will be ultimately be a commercial winner.

When I worked for a defence organisation, this traditionally had operated in a very "closed innovation" environment (due to, for example, defence security concerns of openly describing problems or technologies), I was immediately taken by the principles of "open innovation" and how this might have an impact in the defence market and in the development and pull-through of new materials and technologies.

This epiphany came while I was leading the operation for the Advanced Technology Centre (ATC) at BAE Systems plc, the UK multinational defence, security and aerospace company. The ATC was a traditional research centre with a huge legacy of great people, inventions and capabilities inherited from the old Marconi and British Aerospace organisations, which had been absorbed by BAE Systems. We had a significant number of very clever, capable and knowledgeable scientists and engineers within the parent organisation—but it became clear that there was also great technology and innovation not only in-house but also within our partner universities and from across the wider supply chain. The question I kept asking "if only we were able to access capability in a better way?" So, from 2004, one of the first principles I introduced was to adopt methods on how we might find better ways of engaging with our knowledge partners and how we might bring some of these great ideas together in a more integrated way—using open innovation principles (see Table 1.1).

Any resistance to change was soon mitigated by the realpolitik that our research community then faced. Investment from the defence sector had dropped off and so we had limited R&D budgets flowing down from the parent organisation. As my then colleague Dr John Bagshaw, the ATC's technology executive, told Eureka magazine in June 2012 [2]: "Figures have radically changed. If you go back 15 years and look at the defence R&D budget was about £660m, which was worth about £1.2bn by modern standards. Now the defence R&D budget is about £350m. So you're trying to cover the same sort of degree of technology with less than a third of the funding."

Table 1.1 Contrasting principles of closed and open innovation

Closed innovation principles	Open innovation principles
The smart people in our field work for us.	Not all of the smart people work for us—we need to work with smart people inside **and** outside the organisation.
To profit from R&D we must discover it, develop it and ship it ourselves.	External R&D can create significant value; internal R&D is needed to claim some portion of that value.
If we discover it ourselves, we will get it to the market first.	You do not have to originate the research to benefit from it.
The company that gets innovation to market fist will win.	Building a better business model is better than getting to market first.
If you create the most and best ideas in the industry, we will win.	If you can make the best use of internal and external ideas, we will win.
We should control our IP, so that our competitors do not profit from our ideas,	We should profit from others' use of our IP, and we should buy others' IP whenever its advance our own business model.

Source: H. W. Chesbrough. *Open Innovation: The New Imperative for Creating and Profiting from Technology*, Harvard Business School Press, 2003 [1].

As a result of this funding crisis colleagues quickly realised we could squeeze significantly more from our R&D funding if we adopted a more open innovation approach with our partners and their supply chain. It also became clear that often it was not just the technology alone that makes the key difference for a product or application but more on how we integrate that technology for use in a systems context. This "systems thinking" has been a key factor throughout my career and into my current role at The University of Manchester and with the commercialisation of graphene. It is critical to understand how the technology might be used in a particular area or environment, i.e., from a "market pull" perspective as opposed to a "technology push" type of scenario.

The ability to build better business models by engaging with a broader supply chain and to create value through the commercialisation of technology has been a key principle that drove me throughout my career in industry and more recently being based in academia. This is also a factor in what I believe will be critical for the success in the commercialisation of graphene and other 2D materials.

Later in this book, we will talk about the various research and commercialisation facilities within Manchester's graphene innovation ecosystem. While these centres of excellence are indeed impressive, it is important however that we think about the market perspective when talking about what the facilities do. Rather than thinking "inside out", i.e. what the buildings do, we actually start to think more like "outside in", i.e. what are the market needs and requirements and how do we bring together the right materials manufacturing and product to meet those market demands and what is the role of the facilities in achieving this. With the advancement of the Internet of Things, social media and the ability to share information much more rapidly for new developments and ideas, it is important to establish a model that can bring together the great science not just within The University of Manchester but also all that was taking place in the supply chain both in the UK and broadly on the International stage.

1.1 Early Experience in Open Innovation

Some of my early experiences and examples of open innovation included pioneering research on autonomous systems. When I first started this work in BAE Systems' Advanced Technology Centre (ATC), we initially faced a large number of challenges, both from within the company but also from the customer community. These challenges included the fact, at the times, there were no market requirements for an autonomous system, despite some of the technology futures and foresight activity taking place was pointing to self-driving vehicles.

But then a number of factors were starting to take place in the world of defence, most significantly the move towards "asymmetric warfare" and the fact that in the modern defence climate it became less acceptable for the loss of life or indeed potential capture of personnel by the enemy. For example, a pilot or soldier might be taken prisoner and used for propaganda purposes. So, while there were still a number of stakeholders who could not see the need for autonomous systems, there clearly was a trend emerging towards more autonomy on the battlefield. Therefore, BAE Systems gave the green light for the Advanced Technology Centre to begin a small autonomous vehicle programme which included the development of new software and sensors. What was originally devised as a research platform into autonomous resupply "mules" for carrying equipment in combat zones such as Afghanistan was to prove a seminal project, in terms of both technological advances and innovation delivery.

1.2   The Wildcat Is Let Loose

One of the first capital expenses I authorised in the ATC was the purchase of a Bowler Wildcat, an off-road vehicle derived from the Land Rover Defender platform and built in collaboration with the Bowler specialist racing company based in Derbyshire. Normally these robust machines were used for competing in rallies and off-road racing but for us the Wildcat would literally provide the vehicle for our integration of technologies and for testing.

To develop our autonomous technology we had already decided on an open approach and to create an "open systems" architecture that would host our experimental work. This model would be based around a generic platform that would also allow us to "plug and play" components and software systems that were sourced from the open marketplace, many from commercial applications. For example, we used a readily available collision avoidance system because that technology had matured in the mid-2000s for the automotive "driving-aids" market and we did not want to waste time and resources by reinventing any pre-existing technologies. We just wanted to quickly see how these components would work when put together to achieve autonomous navigation. The Wildcat therefore gave us the platform we needed and within weeks it was adorned with an array of components, sensors and software, all plugged into a single system so we could easily test the arrangement in real-world environments.

The experimental Wildcat was developed in a workshop at our ATC facility in Bristol and became BAE Systems' first full scale autonomous ground vehicle experiment. For several years, we developed software and sensors in the pursuit of autonomous navigation. I described the Wildcat project in some some detail in a blog for the *Huffington Post* [3]:

> "Wildcat started life as a 4×4 off-road production car from Bowler, but was modified by BAE Systems to sense the world around it, plan its own route, navigate and avoid obstacles. This requires the integration and systems engineering of advanced technology including computer controlled steering servos, a secondary braking system, a hotline into the vehicle's engine management system, wireless data links, GPS and laser ranging sensors all coupled to the vehicle's brain where advanced algorithms make intelligent decisions about how to act in the light of the information provided. As a result it cannot just be controlled remotely, but also follow a pre-set path or make fully autonomous decisions about the road ahead and how to navigate obstacles it encounters so that it can complete its journey without any further human involvement."

One of the most memorable experiences for me included following the Wildcat around the MIRA vehicle testing park in the Midlands and watching in horror as a circular component went spinning from the roof of our test car, past our window and landing on the track. To my relief, I was later informed that this rogue disc was in fact a Frisbee that had been carelessly abandoned on the car by a group of student researchers. So, once we had successfully ironed out the various glitches and technical challenges we began to safely demonstrate at various events and conferences where the Wildcat performed live in front of an audience and showcased the capabilities of what autonomy might do for future defence applications.

The car achieved autonomous navigation at 70 km/h and featured autonomous collision avoidance capabilities at 20 to 25 km/h. Not surprisingly, this prototype vehicle was a great way to excite and introduce new students and engineers to the brave new world of autonomous vehicles—indeed, many of these young engineers were inspired to go on to become key members of the R&D community that would work on the various autonomous systems we are starting to see being developed in various markets today

1.3   The Oxford Connection

Even though the development of the autonomous platform was going at a pace, we eventually reached the stage when we realised that we would benefit from a much broader engagement beyond the initial team and the budgets that we had within the ATC. We therefore decided to partner with academia and we donated one of our two experimental vehicles to the University of Oxford as part of a new academic-led programme of work. The university was able to bring, not only new research but also they were able to leverage what was taking place across the commercial sector with the start of a national autonomous vehicle programme that was starting to take shape in the UK. The academic leader was Professor Paul Newman who now heads the Oxford Robotics Institute, which has a global reputation in mobile autonomy and developing machines and robots that can map, navigate and understand their environments. The Oxford group has a focus on transport, including robot autonomy, navigation, mapping, scene understanding and perception. A flagship project is RobotCar UK.

At the time we teamed up with Oxford, we were already starting to see applied developments in the automotive sector, from lane departure systems through to braking systems—all of these being good examples of the first and early stages of an autonomous system. By working in collaboration with Oxford we were able to see significant advances and we were not being held back by the lack of budget and all the cautiousness of the defence sector. Today you can see the significant trends and developments taking place at pace around the whole area of autonomy. Most cars on the road today have some form of an autonomous system within them and in the future we are now starting to see more and more vehicles that have an increased amount of autonomy with their systems. Today, not only can cars perform self-breaking but also boast a range of other autonomously capabilities and decision-making. A great example of this is the autonomous vehicles being developed by that great disrupter of the auto sector, Tesla.

1.4 Bringing the Wildcat Experience to Manchester

The Wildcat project was an early example of my exposure to open innovation principles and today I am very proud to see the significant activity that has taken place since this project was initiated. I believe this breakthrough in technology would have stalled if we had maintained a closed innovation approach on that program and relied exclusively on the limited budget and internal team within BAE Systems.

For me, the Wildcat story was a milestone experience and very influential in what we are looking to achieve with graphene and 2D materials at The University of Manchester. By working and embracing innovation from across academia and industry and looking to create a supply chain in close collaboration with our partners—i.e. pursuing the principle of open innovation—I believe we can really accelerate the adoption of not just graphene but a whole series of advanced materials. That does not mean however that we do not need to create intellectual property in the form of patents or know-how. Again, we have developed innovative business models for the graphene innovation ecosystem at Manchester that enables partnership and collaboration but also protects core intellectual property and know-how and the creation of value through the exploitation of these knowledge assets. However, it is critical, that we accelerate the time taken to create this value in line with the lifespan of the IP. It can take over 10 to 20 years to create value and often the patent expires or becomes less relevant by the end of this development period. The pace of innovation is therefore critical if we are truly to create value from the ideas and inventions in graphene and 2D materials and this will be critical to create value for UK plc.

1.5 Working with SMEs

Another key learning from my industrial career was that of working with small and medium enterprises (SMEs)—an SME in the UK and the EU is generally a business with no more than 250 employees and a turnover no more than €50 million (about £43 million). SMEs make up 99.9% of all businesses in the UK. During my time in BAE Systems, I was the technology representative on a joint UK business/MoD industrial working group. One of my tasks within this network was to chair a working group that brought together the industrial and defence sectors to look at the potential business models that could support a new innovation portal for working between the MoD and small businesses that were traditionally not involved in the defence sector. This initiative was initially called the Centre for Defence Enterprise (CDE) and the group was to look at how we could encourage conservative defence customers to rely less on the well-established big players in the defence supply chain and to consider using other businesses and innovative start-ups that were not currently involved in the defence market.

We supported the development of a novel innovation portal into which new ideas could be presented on a simple "template" to produce rapid prototypes and demonstrators or feasibility studies that could create a benefit for the defence customer. Within the ATC—which was itself effectively acting like an SME—we created special events for brainstorming across different areas of the organisation, and in addition we brought in partners and put forward novel and creative ideas via the Ministry of Defence's innovation portal, now called the Defence and Security Accelerator. We achieved some significant success in a number of these proposals from new materials, through to novel sensors, through to new ideas that could create a benefit for defence.

This was achieved by operating under the open innovation principles but we were also starting to introduce an innovation cycle that featured a rapid "make or break" type of culture (see Fig. 1.1). You could view this approach as if being given permission to fail in developing a new idea but if you were to fail in that endeavour you have to "fail fast" and quickly learn from the experience and move on very quickly. If you are going to create new and great innovations then we should not fear failure but we should really push the boundaries of what we might traditionally look to do but do so not over a long and prolonged timescale. By exploiting modern technology and working across the supply chain and using rapid manufacturing and prototyping techniques, you will be able to demonstrate concepts with rapid prototypes.

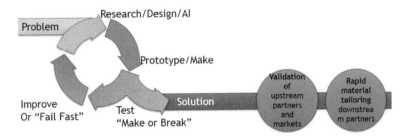

Figure 1.1 Being prepared to fail: the innovation cycle above responds to a problem through research and drawing on digital tools like AI and data to help design a solution. From this work you can develop a prototype (like the Wildcat vehicle) to test and to see if it breaks. If the design does fail then learn quickly from the experience and refine the design ("fail fast") and repeat this cycle until you achieve a reliable solution. The new innovation then needs to pass through various barriers to get to market, i.e. validation and regulation; commercial scale up and working with a supply chain. From James Baker, Liam Britnell, Craig Dawson, ©2020).

These demonstrators could be shown to the customer to achieve immediate feedback and to see whether this has some benefit or utility for defence. Following on from this, a rapid development spiral enables you to significantly reduce the lead time from taking a technology or a concept into the marketplace. This learning also follows the principles from the US and in particular the Defense Advanced Research Projects Agency—or DARPA for short. DARPA is a powerhouse research and development agency that looks to bring often radical ideas rapidly into application. Essentially the agency is never afraid to fail, however exotic some ideas may seem—nor is it bound by the usual resource or procurement challenges due its national strategic importance to the US government.

The concept of "grand challenge" followed by rapid prototyping leads to examples of organisations like the famous Phantom Works the advanced prototyping arm of the defence and security side of Boeing or the Skunkworks (Lockheed Martin) in the US. The Graphene Engineering Innovation Centre (GEIC), a graphene accelerator that we talk about later in this book, was modelled very strongly on the principles from DARPA and this rapid prototyping approach.

This has put the "make or break" philosophy into the heart of the model to commercialise graphene.

References

1. H. W. Chesbrough. *Open Innovation: The New Imperative for Creating and Profiting from Technology*, Harvard Business School Press, 2003.
2. P. Fanning. The cutting edge of design: Inside BAE Systems' Advanced Technology Centre, *Eureka Magazine*, June 14, 2012. https://www.eurekamagazine.co.uk/design-engineering-features/technology/the-cutting-edge-of-design-inside-bae-systems-advanced-technology-centre/43014/.
3. J. Baker. Autonomous vehicles accelerate into the fast lane, *Huffington Post*. https://www.huffingtonpost.co.uk/james-baker/autonomous-vehicles_b_1193430.html.

Chapter 2

Graphene: The New Kid on the Block
By James Baker

My experience in applying open innovation to pioneer autonomous driving and navigation systems and the advocacy of "fail fast, learn fast" would prove invaluable as I found myself taking up a new role at The University of Manchester working on graphene commercialisation. It was now 2013 and graphene, the world's first two-dimensional material, was less than 10 years old and still in the very early stages of technology and academic development. However, at this time there was already some fantastic discussion and dialogue around the potential products and applications.

The application that immediately hit my emotional interests was the ability to take dirty, salty water through a graphene membrane to create clean drinking water. I immediately imagined this vision, of a graphene straw that allowed users to drink water directly from the sea—a device that could change the lives not only of individuals but of countries and regions across the world. How could I not want to be involved in something of that potential which truly could make a difference, not just to the business world but to people across the globe? I was very therefore very excited to join The University of Manchester in this exciting role and responsible for developing the partnerships and collaborations with industry to commercialise this new material.

Graphene: The Route to Commercialisation
James Baker and James Tallentire
Copyright © 2022 Jenny Stanford Publishing Pte. Ltd.
ISBN 978-981-4877-87-9 (Hardcover), 978-1-003-20027-7 (eBook)
www.jennystanford.com

This anticipation was heightened by the fact that, in my previous role running the Advanced Technology Centre (ATC) at BAE Systems, I had in fact interacted with The University of Manchester. In spite of the open innovation approach at the ATC, frustratingly I had found only a very small amount of time to fully investigate what possibilities lay within the great university sector across the UK. Even though I had worked with The University of Manchester on previous occasions, including some visits to see some academics and facilities, I had not fully appreciated the depth and breadth of capability and talent in science and technology that was spread across this world-leading institution.

So, with a lot of enthusiasm and some knowledge of the UK higher education sector, I came to Manchester with the challenge to help bring this so-called "wonder material" to market. What was particularly exciting was the fact I was coming to the place of graphene's original discovery and there was a real commitment from the University, and indeed the wider city-region, to make sure Manchester would lead this material's journey to commercialisation. There was genuine pride that Manchester was the "home of graphene" through research and discovery but needed to also become the home of graphene through commercialisation.

2.1 The Accidental Pioneers

So in all the cities, in all the world, why was graphene discovered in Manchester? Well, it was open curiosity that led to the isolation of the world's first two-dimensional material. Andre Geim, a Russian-born, Dutch-British physicist was working with fellow Russian émigré Konstantin (Kostya) Novoselov when, almost by accident, they realised that they had managed to reduce graphite to a single atomic layer which possessed amazing capabilities (see Fact File below). Andre was supervising Kostya during the completion of his PhD. To satisfy their almost playful curiosity the pair held regular "Friday night experiments" where they would try out experimental science that was not necessarily linked to their day jobs. Kostya makes clear that the special approach taken by his former mentor Andre Geim was fundamental in the isolation of graphene—a methodology not too dissimilar to the "fail fast, learn fast" innovation philosophy described in Chapter One. While Andre is keen to emphasise that the breakthrough was always a team effort that involved other colleagues. These included Professor Irina Grigorieva and Sergei Morozov, who are still connected to The University of Manchester. "While several others are scattered around the world in academia and industry," added Andre.

Graphene's origin story is described in detail by Kostya in a special series of talks organised by The University of Manchester during the summer of 2020 entitled the "Lockdown Lectures" [1]. Kostya was interviewed by Manchester third-year politics student Megan Ritchie, and during their conversation the renowned physicist was asked how the isolation of graphene had come about. Kostya explained: "The most important teacher in my life is Andre Geim with whom I started to do my PhD in Holland some time ago.

"Andre always had, and has up to now, this interesting way of doing science through small projects, which we called the "Friday night experiments"—and it is really enjoyable working with him. Andre is known for his levitation of frogs, we had also been doing gecko tape together and "magnetic water"; so graphene indeed was one of the Friday evening experiments when you ask a question and try and resolve it with minimum effort spent. If it works that is great, if it does not work you have not wasted too much time.

"Graphene was one of those experiments. We tried, it failed, and we almost forget about this—[but] the story was, I was present in the lab when I saw people cleaning graphite to use in the scanning tunnelling microscope as a sample. I knew about this, when they take graphite and clean the top with this Scotch tape and throw the Scotch tape away. I have seen it many times before but it is only when your brain is thinking about how to make graphene and you put two and two together and you pick up this Scotch tape from the dustbin and make [an experimental] transistor [device] out of it.

"We made our first device very quickly. Within half an-hour of picking up that Scotch tape we had made our fist device; it wasn't graphene at the time, it was still a thin layer of graphite but its properties were such we knew something interesting was going on.

"I should say when we started to do this we knew for sure, or at least I knew for sure, that graphene *shouldn't exist,* so we were making transistors out of graphite and we were enjoying studying their properties. Then when we made devices which showed some very special performance and we couldn't explain their behaviour [unless we had] anything but exactly a monolayer of graphene—then we really realised we had made this one atomic thick material.

"It took us about a year to get to our first [graphene] devices and we were very lucky that we had a device with one-layer and two-layers of graphene and they behaved completely differently, that was how we knew [single layer graphene had unique properties].

"I guess you can say it was an accident and accidents are important in science first of all, but also accidents never happen accidently, you actually need to create an environment for those accidents to happen; that's the difference between and good scientist and a bad scientist. So, the good scientists create the environment for as many as possible of those accidents happening."

2.2 Playfulness Rewarded with a Nobel Prize

In October 2004, Andre and Kostya co-published a paper announcing the achievement of graphene sheets in *Science* magazine, entitled "Electric field effect in atomically thin carbon films" [2]. It is now one of the most highly cited papers in materials physics and by 2005 researchers had succeeded in isolating graphene sheets—so sparking a revolution in materials science which still has an impact today. Interest and investment in graphene research was to be exponential, a science "gold rush" (see Chapter 11). The extraordinary graphene phenomenon was to lead to Andre and Kostya being awarded the Nobel Prize for Physics in 2010. In the official press pack released on October 5, 2010, the Royal Swedish Academy of Sciences announced it "... has decided to award the Nobel Prize in Physics for 2010 to Andre Geim and Konstantin Novoselov, *'for groundbreaking experiments regarding the two-dimensional material graphene'*". In direct reference to Andre and Kostya's approach to experimentation, the academy also said: "Playfulness is one of their hallmarks. With the building blocks they have at their disposal they attempt to create something new, sometimes even by just allowing their brains to meander aimlessly. One always learns something in the process and, who knows, you may even hit the jackpot. Like now, when with graphene, Andre Geim and Konstantin Novoselov have written themselves into the annals of science" [3].

Figure 2.1 Andre Geim officially receives his Nobel Prize for Physics at the official Prize Award Ceremony in 2010 hosted in Stockholm, Sweden. Image credit: Janerik Henriksson/Scanpix. https://www.nobelprize.org/prizes/physics/2010/geim/photo-gallery/.

2.3 Graphene Becomes a Global Icon

After the Nobel Prize, Andre and Kostya were further rewarded by a grateful nation. The two Russian-born scientists were knighted in 2011 and then gifted the Freedom of the City of Manchester in 2014. And this recognition of Manchester as the mecca for graphene helped attract some impressive visitors, including China's President Xi Jinping, who witnessed the University's graphene facilities first hand as a finale to his state visit to the UK in 2015; the next year, the

Duke and Duchess of Cambridge followed in the Chinese premiere's footsteps on a similar fact-finding tour. As well as the public acclaim and VIP visits, academic interest in the science around graphene also took off like a rocket. "It can be stated without hesitation that [the] Nobel Prize emphasised the importance of this subject and accelerated research in the area," said George Wypych in "Graphene: Important Results and Applications" (2019) [4].

> **FACT FILE: Graphene and the 2D material family**
>
> Graphene is the name for a honeycomb sheet of carbon atoms. It is the building block of graphite—as found in pencil lead—and is an incredibly strong and conductive material. It can be used in a wide range of applications, from aerospace engineering to digital electronics and biomedicine.
>
> At nanoscale, graphene is the strongest known material, yet it is also stretchy. It can conduct electricity and heat incredibly well, for example its high electron mobility is 100× faster than silicon and its electrical conductivity is 13× better than copper and graphene conducts heat 2× better than diamond*.
>
> Graphene can absorb only 2.3% of reflecting light; it is impervious so that even the smallest atom (helium) cannot pass through a defect-free monolayer graphene sheet; and its high surface area of 2,630 square meters per gram means that with less than 3 grams you could cover an entire soccer field*.
>
> The isolation of graphene at Manchester led to the discovery of a whole family of 2D materials and crystals, including hexagonal boron nitride and molybdenum disulphide. These "brothers and sisters of graphene" have been dubbed "the Graphenes".
>
> Once combined these layered 2D materials can create new "designer materials" known as heterostructures to deliver an unprecedented range of novel materials and multi-functional materials. Essentially, this means new materials can be built from the ground up on an atomic level to create materials tailored to exact functions.
>
> *Characteristics reported in *Nanowerk* 2019 [5].

Graphene was to be one the greatest scientific stories of the century so far. But its commercial potential was, in fact, also recognised early on. My Manchester colleague Alan Ferns, the Univerisity's former Associate President for External Relations and Reputation, has revealed that at quite an early stage the University's leadership and it stakeholders began to see the potential of graphene "as a kind of commercialisation opportunity, an iconic opportunity" that could persuade investors from Europe, Asia and the USA to look to Manchester for the R&D expertise to help support new market opportunities around graphene.

It is this ambition that created what Alan Ferns describes as an "end goal" and inspired the direction of Manchester's graphene narrative and engagement strategies to date. To help deliver on this hugely ambitious mission the University went about leveraging support from high-level sponsors, including the British government, in a bid to attract funding to Manchester which would be used to help build not one but two facilities to help take graphene discovery from lab to market. My new role was to help manage this commercialisation process and be one of the architects of a new graphene city.

References

1. The University of Manchester's Lockdown Lecture featuring Kostya Novoselov, May 27, 2020. https://www.manchester.ac.uk/coronavirus-response/coronavirus-home-learning/lockdown-lectures/kostya-novoselov/.
2. K. S. Novoselov, A. K. Geim, S. V. Morozov, D. Jiang, Y. Zhang, S. V. Dubonos, I. V. Grigorieva, A. A. Firsov. Electric field effect in atomically thin carbon films, Science, 2004, 306(5696), 666–669. https://science.sciencemag.org/content/306/5696/666.full.
3. Nobel Poster from the Nobel Committee for Physics, web adapted by Nobelprize.org   https://www.nobelprize.org/prizes/physics/2010/illustrated-information/.
4. G. Wypych. *Graphene: Important Results and Applications*, 2019, ChemTec Publishing.
5. Nanowerk Graphene Description. https://www.nanowerk.com/what_is_graphene.php#:~:text=Graphene%20possesses%20other%20amazing%20characteristics,helium)%20can't%20pass%20through.

Chapter 3

Graphene's Great Expectations
By James Baker

When I originally heard of graphene, it was often associated with a list of superlatives in association with its capabilities. For example, boasting properties of being 200 times stronger than steel (see Fig. 3.1), more conductive than copper, flexible, transparent and having the ability to allow certain molecules to pass through but to block others. As a shorthand, journalists simply dubbed graphene the "wonder material".

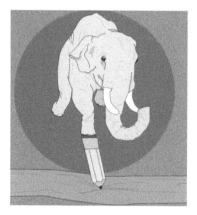

Figure 3.1 Super strong: a metaphoric representation of a graphene sheet suspending the full weight of an elephant balancing on a pencil. Graphic supplied by Ella Byworth. https://www.ellabyworth.com/.

Graphene: The Route to Commercialisation
James Baker and James Tallentire
Copyright © 2022 Jenny Stanford Publishing Pte. Ltd.
ISBN 978-981-4877-87-9 (Hardcover), 978-1-003-20027-7 (eBook)
www.jennystanford.com

Such labels seemed only to raise the expectations of graphene's potential applications, which seemed endless. As mentioned, an area that made me immediately excited was that of its filtration properties and having the ability to allow certain molecules to pass through but to block others. This makes you think about the ability to create clean drinking water from dirty salty contaminated water—and this would not only create a huge commercial opportunity but also have the ability to transform the way we obtain and use water across the world.

The list of applications continues from biomedical applications, in terms of artificial skin to smart bandages, through to drug delivery and ultimatley new neurological treatments to help tackle brain disorders such as epilepsy and Parkinson's Disease are possible, and papers have been published about these breakthroughs within the medical sector. You can also start to see how graphene might be used for example around the environmental and sustainability agenda from replacement or reduction in plastics through to new food packaging featuring smart sensors that tells you when the contents are starting to deteriorate or to go bad. Or in energy applications, the ability to make batteries last longer or to create a new form of battery or supercapacitor that can be rapidly charged and reused over numerous occasions.

However, perhaps because of my background in the defence industry, an obvious sector that could be transformed was aerospace—a sector that is on the constant quest to produce lighter but stronger aircraft which therefore have greater range using less fuel and with lower emissions. To achieve this more sustainable vision, one of the first challenges the sector has been looking to improve over many years is that of lightweigting. The very first aircraft were made from wood and these were followed by aircraft made of metals, with aluminium being a principal material as well as more exotic materials like titanium.

3.1 "Plastic Planes": They Will Never Take Off!

In the 1960s, we saw the discovery of carbon fibre and many people at that time were very dubious whether this new material would ever find its way into an aircraft of the future. From my background in aerospace I was aware of some great stories and influential characters in the industry who were very sceptical that a "plastic aircraft" would ever find its way into mass production. If you look however at modern aircraft, you will see a significant number of planes in modern fleets featuring a significant proportion of carbon fibre structure within them. There continues to be a push towards lighter, quieter, greener and more fuel-efficient aircraft to meet the needs and expectations of society today. The Covid-19 crisis and its devastating impact on the travel industry only adds to this push for more innovation to develop aircraft for the "new normal" of commercial flight.

So, imagine if we can add graphene to the existing carbon fibre structure. This new composite would, first of all, reduce the amounts of carbon fibre material need to make a plane so bringing down manufacture costs; but the graphene-enhanced carbon fibre would also reduce weight of the plane itself and therefore lower the cost of flying for any aircraft fleet. If graphene could enable this, then clearly this has a role for our aircraft of the future. Where graphene starts to get incredibly exciting is around multifunctionality. What if, not only in the lightweighting of a wing or a structure but imagine if we could also use graphene to transmit heat across the wing surface to achieve de-icing of that wing. Again, we could not only reduce weight and cost but also improve the efficiency and performance of our aircraft.

It does not stop there, however. Aircraft today also must withstand a direct strike from lightning onto their structure in flight and survive this significant effect. This is something that if you are unlucky to experience can be quite a scary incident. However, you might be very pleased to discover that aircraft today are designed and manufactured to withstand such a lightning strike. This is achieved either by applying specialist metal components or by embedding metal, for example copper mesh, into the composite structure so that the aircraft will form a Faraday cage in such a lightning strike situation.

If we can exploit the electrical conductivity properties of graphene either to enhance or to improve that structure, again this has opportunity to reduce weight and improve performance and efficiency of aircraft of the future. As we continue to think about how we might improve next-generation aircraft, graphene could also acts as a strain sensor which we could build into the structure as a "health management system" for the wing informing operators if damage has occurred during flight or maintenance; and this warning system could also mean we have an increased confidence with planned maintenance of our aircraft.

If you can continue with this imagination of possibilities, you can start to see how graphene could make a game-changing impact in aerospace.

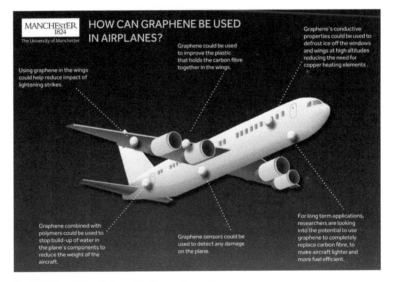

Figure 3.2 Potential applications of graphene in aerospace.

3.2 World's First Graphene Plane

A chance encounter helped me to put some of these ideas into action. I had only been at The University of Manchester for less than one year when I had been asked to speak about graphene at a conference on advanced manufacturing and materials in the Greater Manchester area. I had just finished my presentation when I was approached at the break by a familiar face from my past, Billy Beggs, who was an ex-colleague from BAE Systems.

Billy had seen that I was speaking at the conference and had wondered about this "wonder material" called graphene and how it might have an impact on the aerospace sector. Following our conversation, we decided to collaborate and to see how we might accelerate the adoption of graphene for the aerospace sector. Billy having left BAE Systems had joined the University of Central Lancaster (UCLan) but was also in touch with several small businesses and ex-colleagues involved in the manufacture of Unmanned Ariel Vehicles (UAVs). We decided that by working collaboratively together we could explore the benefits graphene could potentially bring across aerospace, from making aircraft stronger and lighter through to improvements in energy storage and even in enhancing the vehicle's health management. Billy had significant achievements around technology integration and he has achieved first flights of a series of pioneering platforms you see flying today; Billy was keen to see how we might achieve a first flight of an aircraft using a graphene component, initially on a small UAV. Unfortunately, as many have found out, you just cannot make a wing out of graphene and enable it to fly very easily. So, after several conversations and challenges from Billy we then managed to get a group of companies together and within a 12-month period we achieved a first flight of a graphene enhanced UAV from a field in Lancashire.

This first flight was very basic and it was essentially a wing painted with graphene and arguably the graphene did very little in terms of functionality although we did find some positive results in terms of airflow which the graphene might have had a role in enhancing. However, this very early test machine did demonstrate how to create a team of people from academia and industry and a supply chain that is able to very rapidly go on to produce the first prototype graphene enhanced UAV that featured graphene enhanced carbon fibre. And although this was still a very basic integration, the graphene developments were now starting to move with some pace. A phase two prototype demonstrated that the supply chain was moving from just making a few grammes of graphene material into manufacturing a real graphene enhanced component at a size that could be integrated into the wing of a small aircraft.

To showcase this success, we flew this prototype at the Farnborough air show in 2016. This experimental programme showed that by bringing together the right people from across academia and industry to resolve a real world challenge we were able to very rapidly produce a working demonstrator or prototype that achieved a world first flight in front of tens of thousands of spectators. Clearly this was a calculated risk and thanks to the great skills of Billy and the team was a great success. The flight also achieved a significant amount of interest from the aerospace industry and community and we have continued the development of the UAV and other components for aerospace and further flights of more capable, next-generation vehicles will continue as we go forward.

3.3   Richard Branson Backs Graphene

Our experiments in aerospace were also successful in highlighting graphene's potential to Sir Richard Branson, the founder of Virgin Atlantic, who would go on to be a very high-profile champion for graphene's adoption by the aerospace sector. Even before the Covid-19 crisis and its catastrophic impact on the air travel industry, Sir Richard had already recognised the sector was facing economic and environmental pressures to radically restructure.

Sir Richard went public on the need to adopt 2D materials, including a much-reported speech in Seattle in 2017. With such an influential voice supporting graphene's disruptive potential in the aviation industry, we went on to produce a strategy paper in 2018 [1] with the Aerospace Technology Institute (ATI), which is responsible for the technology strategy for the UK aerospace sector. This report was endorsed by Sir Richard, himself a previous champion of carbon fibre and an advocate for the aircraft industry to reduce fuel and emissions to improve the performance of both current and future fleets of aircraft. Sir Richard said: "The potential for graphene to solve enduring challenges within the aerospace sector presents real opportunities for the material to become disruptive, and a key enabler in future aircraft technology. We need to accelerate the opportunity for the UK to realise the benefits from graphene by creating a portfolio of graphene related research and technology projects which if undertaken would lead to real impact in our aerospace industry."

The ATI-Manchester joint INSIGHTS paper [1] which was produced in consultation with a range of stakeholders and the subsequent report highlighted the potential of graphene to positively impact on aircraft performance, cost and fuel efficiency. For example, the safety and performance properties of aircraft could also be significantly improved by working with the existing supply chain to incorporate graphene into materials already used to build planes. From this strategic vision, the ATI has pledged to accelerate the maturation of graphene technology opportunities through its research and technology programme. Again, the collaboration between industry and academia was cited as being crucial in the R&D ambition.

3.4 Graphene Drives Auto Innovation

Another sector that is facing massive restructuring is the auto industry. Again, this call for change was being made long before the global Covid-19 pandemic, although that crisis has only added new pressures to accelerate innovation.

In a bid to meet its net zero ambitions, the UK government has ruled that as a nation we will have to stop buying petrol and diesel engine vehicles by 2035. This has therefore resulted in a consumer push towards electric powered vehicles and as a result there has been significant developments, driven initially by pioneering electric-car maker Tesla in the US but today adopted almost universally by all automotive manufacturers.

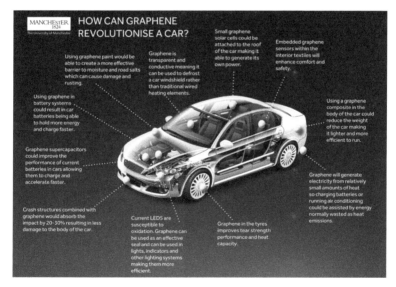

Figure 3.3 Potential applications of graphene in automotive.

The challenge that the auto industry shares with the aerospace sector is the need to address the light weighting of vehicles. Because of the increased weight of their large battery powertrain, electric cars are heavier than traditional vehicles using internal combustion engines which means e-cars will have an impact on our roads and infrastructure. So not only do we need to investigate new forms of electric vehicles and perhaps the battery and supercapacitor hybrid is something that of could be of great interest—but we also need to think more of the system, i.e. how we make the overall vehicle lighter but also how we optimise our roads and infrastructure in the future.

3.5 Better Batteries, Supercapacitors and Hybrids

If we begin by looking at the battery, by putting graphene into the anode cathode or even the membrane then potentially 2D materials can play a role in terms of enhancing the life and performance of our batteries in the future. Graphene and the some of the Graphenes—i.e. the growing family of other 2D materials and their application to create new technologies and products—could also play a role in terms of new types of batteries and energy storage devices. For example, graphene and 2D materials could play a key role in developing supercapacitors, a new generation of energy storage devices that are expected to deliver rapid charge or discharge cycles, offer robustness over multiple cycles, while providing safe operation at minimum weight. In the future, I believe we will start to see hybrid structures whereby our cars are not only containing batteries to power our cars but they will be complemented by a supercapacitor that will be integrated into the structure of the new e-vehicles. So a new, hybrid vehicle, with an integrated battery structural framework, could start to really achieve the objectives and demands of our future automotive sector.

3.6 World's First "Graphene-Bodied" Car

The other major offer graphene can make to carmakers is its lightweight capability and here we are seeing some early breakthroughs. Briggs Automotive Company (BAC), the Liverpool-based maker of the BAC Mono single-seater road-legal sports car, has put graphene into its vehicle. Its newest model has been hailed as "the first production car in the world to fully incorporate the use of graphene-enhanced carbon fibre in everybody panel". Graphene has been used to enhance the structural properties of the fibre to make panels both stronger and lighter, while also significantly improving the car's mechanical and thermal performance. This world-first for BAC followed a successful Advanced Propulsion Centre-funded research and development project into the production-readiness of graphene. The brand, working alongside Haydale—a partner with The University of Manchester's Graphene Engineering Innovation Centre (GEIC)—and Pentaxia, through the Niche Vehicle Network (NVN). What is worth noting from this pioneering project, you only need a very small amount of graphene to make a big difference. Probably less than 1% graphene must be put into a resin or into a foam or into a rubber to make a double-digit benefit. So, BAC is taking significant weight out of the car from just a very, very small amount of graphene.

3.7 The Covid-19 Catalyst

As mentioned, one of the most unexpected and perhaps most impactful innovation drivers since the Second World War has been the Covid-19 crisis. The coronavirus pandemic decimated international air travel in 2020 and sent shockwaves throughout the whole aerospace industry which, by the autumn of 2020, was being forced into making huge layoffs. The situation is just as challenging for many carmakers across the world that have also seen demand in their markets dramatically drop. These industries, whose supply chains often overlap, now have a make or break opportunity to use the Covid-19 challenge as a catalyst to accelerate the innovation cycle—and in doing so, look to become more sustainable and greener sectors. As described, graphene and 2D materials can play a key role in supporting this huge ambition—but manufacturers and their respective supply chains will have to operate very differently than they have done traditionally.

3.8 Fast Lane and the Slow Lane

The experimentation with partners in the aerospace and auto sectors has revealed two approaches to graphene commercialisation.

It can still take many years for a totally new material to reach the marketplace, however by the addition of graphene to an existing material then benefits and applications can be found very quickly. So, we almost have two swim-lanes of innovation; a "slow swim lane" for bespoke applications using the unique properties of graphene and a "fast swim lane" of linear innovation, in that by adding a very, very small amount of graphene to an existing material you can start to transform the properties of that material. This fast lane approach is more likely to be seen in existing supply chains as manufacturers do not have to radically restructure to rapidly pull innovation through into their products.

3.9   In the Fast Lane: The Rubber Experience

So if we focus on the "fast swim lane" for the moment there are many products and applications that can be enhanced by the addition of a relatively small amount of graphene to the existing mix or formulation of that material. This is of importance in that often people assume that the addition of graphene to a material would increase the cost of that material. What we are now starting to see is, not only by adding graphene to an existing material can we improve the final performance of that product, you are also finding that you can use less of the original material. So, in fact you can reduce the material cost or increase the productivity of your existing product or material. Some of the early examples we have seen have been with rubber, whereby you can achieve the same performance by using up to 33% to 50% less rubber material by the addition of graphene. This in turn has the effect of a reduction in material cost; while you can also you can also achieve a reduction in manufacturing costs, for example, less energy costs through a reduction in cure times, which has significant benefit across a number of sectors in terms of throughput and product productivity.

The other factor that is starting to be seen by the addition of graphene is that is what we refer to as multifunctionality. An example of this with rubber is that not only can you improve the mechanical strength you might also be able to achieve an improvement in its thermal or electrical conductivity performance. Going even further you might also find that there are other properties of graphene that further enhance the product, a good example might be that of flame retardancy or noise reduction. The Ford Motor Company in now adding graphene to foams and other under the bonnet components. The US carmaker, in collaboration with Eagle Industries and XG Sciences, has found a way to use small amounts in fuel rail covers, pump covers and front engine covers to optimise their performance. Generally, attempting to reduce noise inside vehicle cabins means adding more material and weight, but with graphene, the opposite is achieved. As Debbie Mielewski, Ford senior technical leader, sustainability and emerging materials, says: "The breakthrough here is not in the material, but in how we are using it. We are able to use a very small amount, less than a half percent, to help us achieve significant enhancements in durability, sound resistance and weight reduction—applications that others have not focused on." [2]

3.10 Keep on Running: The World's First Graphene Shoes

One of the earliest commercialisation milestones for Manchester was, in fact, involving a rubber composite. This was the fruit of a partnership with UK-based performance sportswear company inov-8, which worked with a team of composite specialists led by Dr Aravind Vijayaraghavan, Professor in Nanomaterials at The University of Manchester. The collaboration produced the world's first-ever sports shoes to utilise graphene after developing rubber outsoles for a range of running and fitness shoes that in testing have outlasted 1,000 miles and are scientifically proven to be 50% harder wearing. After being launched in 2018, the G-SERIES, which is sold worldwide, has gone on to be a big success for inov-8. One of the most significant aspects of this case study was the speed of getting this product from initial discussions to market, a period of about 18 months. This expediency was due largely to the fact inov-8 is a relatively young brand with lots of energy to bring innovation to its products as part of its brand promise. Aravind is himself an entrepreneur who is now leading his own spin-out interests (see Innovation Case Study below).

INNOVATION CASE STUDY
Keep on running: developing the world's first graphene shoes

Working with inov-8, a specialist sportswear company based in the north-west of England, The University of Manchester has helped to develop the world's first graphene sports shoes.

The project has set the bar high, resulting in not only a world-leading product, but also a highly effective partnership that is boosting the University's commercial reputation—and that of a fellow northern brand.

Birth of the G-SERIES

In 2016 a team headed by Dr Aravind Vijayaraghavan at Manchester's National Graphene Institute (NGI) published a paper revealing how the mechanical properties of rubber could be dramatically improved by adding graphene to it. The story grabbed the attention of the press—and, subsequently, of Ian Bailey, CEO at Cumbria-based sports brand inov-8, who quickly picked up the phone and called Dr Vijayaraghavan to discuss the possibility of collaboration. Following one meeting and two successful funding applications to Manchester's Impact Accelerator Fund and the Knowledge Transfer Partnership (KTP) scheme plans to develop the G-SERIES—new running and fitness shoes with graphene infused soles—were set in motion in early 2017.

Northern innovation

"Everything we do is about innovation and grip," says Michael Price, inov-8's Product and Marketing Director. "We want to be the reference brand—the best brand for committed sportspeople in running and fitness.

"Such sportspeople use their shoes in tough conditions: ultra-marathons, running up and down fells, and so on. Grip is therefore a vital part of the shoe's performance, but high-grip rubber is soft and can wear down more quickly." Graphene@Manchester scientists worked closely with inov-8's factory and field testers to tackle this trade-off, infusing rubber with graphene to create an innovative composite that made the outsoles of the G-SERIES 50% stronger, 50% more elastic and 50% harder wearing. Available in 250 retailers across 26 countries, the new G-SERIES have proved hugely popular.

The need for speed

With 28 years' experience in the sports industry, including at global brands such as Adidas, Slazenger and Asics, Michael knows how important it is to stay one step ahead of the competition. Speed was therefore a key benefit of the project for him—not only in terms of the outcomes for runners, but also in the product's rapid transition from idea to delivery.

"For a sports brand to go through the entire development process and get a brand-new product to market in 18 months is very fast; it usually takes at least three years," Michael explains. "I think this is unique to inov-8 and the University. We've formed a very strong partnership and worked very quickly together."

For Aravind, it was the combination of Manchester's quality of science and support with inov-8's status as a forward thinking business that resulted in this fast-paced outcome. He said: "The science base of graphene at Manchester is the best anywhere by a long way. The University has really invested in commercialisation. Plus, as a small company, inov-8 is very agile and open. We were able to work directly with their factory from a very early stage, trying out compounds on a pilot scale quickly and effectively."

Future developments

The inov-8 KTP was set to run until 2020—but Michael is keen for the relationship to continue further.

"This will hopefully become a long-term partnership," he says. "We're already looking at the next innovation. Graphene is super-light and super-strong, so we can reduce product weight while maintaining strong performance. There's also process innovation; how we create the composites. We're leading the way and will continue to push our leadership position. It's very exciting."

NOTE: An abridged version of an article first published in The University of Manchester magazine [3]

3.11 Need for a New Model Approach

Graphene and 2D materials promise to transform a diverse range of industries by bringing nanoscale capabilities into the real world of 3D products. This has never been possible before in human history. In this chapter I have focused on two manufacturing behemoths—aerospace and automotive—because I believe advanced materials could really help each of these sectors achieve the technological transformation they need to be fit-for-purpose in a fast-shifting marketplace. A revolutionary technology always starts as "new science" that is pulled through an innovation cycle to get it adopted into a market-ready product. To date, with graphene, this has worked for high-performance products, such as the inov-8 "graphene shoe" and the BAC Mono car; or the expectation is we enhance existing components until a tipping point is reached and materials innovation translates across the whole supply chain and into the end-product.

But, in my view, this may not be good enough in response to today's tumultuous times with global challenges posed by Covid-19 and climate change. The innovation cycle needs to be completed at an unprecedented pace—look at how we have delivered a number of novel anti-Covid-19 vaccines in months rather than years. In a similar way we need to put more focus on the transformational potential that bespoke graphene-based technology can delivery but would normally take years in a linear "slow lane". This disruptive innovation also needs to be switched into the "fast lane".

A new approach is needed to deliver this accelerated response and to push innovation much faster in the traditional "slow lane"—and the next chapter will describe how this model, fit for our times, has already been achieved in Manchester, the birthplace of the first industrial revolution.

References

1. Graphene Exploitation: Material applications in aerospace, INSIGHT report, co-authored by the Aerospace Technology Institute (ATI) and The University of Manchester. https://www.ati.org.uk/media/xtijkkrk/ati-insight_06-graphene-exploitation_materials-applications-in-aerospace.pdf.

2. Cell phones, sporting goods, and saloon cars: Ford innovates with 'miracle' material, powerful graphene for cars, Ford Motor Company press release, October 9, 2018. https://media.ford.com/content/fordmedia/fna/us/en/news/2018/10/09/ford-innovates-with-miracle-material-powerful-graphene-for-vehicle-parts.html.

3. A runaway success, *The University of Manchester Magazine*, Autumn 2018 edition. https://www.manchester.ac.uk/discover/magazine/features/runaway-success/.

Chapter 4

Overcoming the Challenge: A New Model Approach

By James Baker

The UK has a reputation for doing great science but has not always been able to take full advantage of the translation of that invention into applications, products and ultimately jobs. Often that value goes overseas to our competitors who create efficient supply chain and highly advanced manufacturing centres that employ high skilled, high-wage workers. A classic example is the development of the world's first programmable computer at The University of Manchester following the explosion of British invention in the wake of the Second World War. Yet, like so many other technological advances that originated in the UK, Manchester saw it world's leadership in computing being eroded and then lost to bigger and better rivals in the USA and then Japan (see Chapter 10).

It became clear to me after arriving at The University of Manchester there was determination to learn lessons from the past and ensure graphene's future was to be anchored in Manchester with a radical model to commercialise a newly discovered material. However, if you look into the historic development cycles of previous new materials, it is often the case that it can take decades from the first discovery of a new material through to the creation of the first products and applications hitting the marketplace. It is that

---

Graphene: The Route to Commercialisation
James Baker and James Tallentire
Copyright © 2022 Jenny Stanford Publishing Pte. Ltd.
ISBN 978-981-4877-87-9 (Hardcover), 978-1-003-20027-7 (eBook)
www.jennystanford.com

timescale that has often been the cause of the science being led in the UK but the supply chain being created by overseas countries—and often this is due to other industries and countries having different business or investment models for taking science into products and applications.

4.1 The Innovation Journey: The Long and Winding Road

In particular, in recent years we have seen an increase of this innovation push taking place in countries like China, which, while perhaps not having the greatest reputation for fundamental science and invention, often very quickly translate new ideas into products and applications, supported with the rapid creation of factories and supply chain companies. If we look at the many case studies around new materials development, with good examples being silicon and more recently carbon fibre, we could argue both are great examples of the classic timeline of going from discovery through to the first products hitting the marketplace taking around 25 to 30 years. Carbon fibre, as an example, also follows the trend of other new materials in that these early products tend to be in niche sporting goods, like tennis rackets or Formula 1 racing car components, then eventually through to defence and aerospace applications. Even now, 60 years or so since its first discovery, carbon fibre is only just beginning to be adopted widely across aerospace and it is only starting to find its way into the mainstream automotive and adjacent markets.

This timescale is due not only to the fact that the actual application of a new technology needs to be addressed but also, we need to understand the other various manufacturing processes, equipment and methodologies that are also required to translate a material into products and applications. For example, in order to take a product into the marketplace you must consider compliance with the various regulation, certification and other legal caveats. In the case of aerospace, the regulation and certification can itself take many decades and, if you do not fully understand the material characterisation and performance criteria, then this could be a critical factor in any delay in accessing a market.

Following the Nobel Prize being awarded in 2010 to graphene pioneers Andre Geim and Kostya Novoselov, industry started to take note of this new kid on the block—a "wonder material" that was associated with a string of superlative properties that could be adopted across a range of products and applications to potentially disrupt many different markets. Graphene has started its route to commercialisation.

4.2 Getting Past the Hype

Industry leaders were quite rightly keen to understand more about how they might use this material and they became initially quite excited by the prospect of adopting graphene into their product range. However, if you have studied the Gartner Innovation Cycle (see Fig. 4.1) then you realise that sometimes you need to go through different phases of a discovery or invention—or, in this case, a new material development cycle—before any real benefit is realised. This often leads to what is referred to as the "hype curve", and clearly graphene in the early days suffered from promising great expectations, which for many only resulted in disappointment. It was soon found that graphene was not so easy to adopt into products and applications as some people felt it might be.

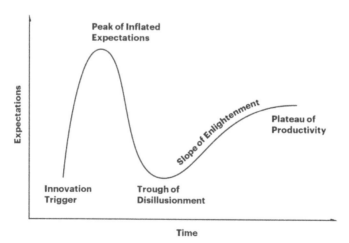

Figure 4.1 The Gartner Hype Cycle. From www.gartner.com/en/research/methodologies/gartner-hype-cycle.

4.3 Electronics Possibilities

For example, there were many early case studies looking at adding graphene into the mobile phone products and there was much early talk around the potential for mobile phones of the future using graphene. These future benefits included phones that not only featured touch sensitive and conductive screens that were tough and less resistance to breakage, but they might also be fully flexible. It was predicted that not only could you have your phone in your pocket you could wrap it around your wrist, you could stitch it into your clothing or you could fold it up into smaller sections and into your pocket like a handkerchief. While all these examples were quite exciting, there are still significant challenges in realising the ambition and producing such sci-fi devices at scale. I am confident that there is potential to see some very innovative developments in the future—but more near-term you are starting to see graphene added into coatings or into pastes or inks doing a much more simple thermal management role in electronic products and applications.

4.4 Technology Readiness versus System Readiness

To get beyond all the hype we therefore need to step back a little and look at the bigger picture. In order to take a new product from discovery through to commercialisation there are many steps and phases that are needed to be undertaken. Often people discuss this in terms of the technology readiness level (TRL)—a process first developed and used by the NASA scientists to understand the various stages and maturity of technology development. The TRL scale goes from initial first discovery, which is usually classed as TRL1, through to a mature technology in the actual representative field of use, which is awarded the highest classification of TRL9.

During my career at BAE Systems, we also believed it was important to adopt a second access and not only look at just the technology issues in taking a new material to market. We decided to also use a system readiness level (SRL) as a second axis to capture not only the technology issues but also the various manufacturing, commercial and other considerations, such as regulatory compliance that can provide barriers to accessing a market (see Fig. 4.2).

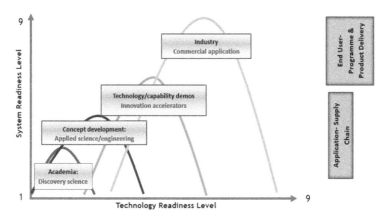

Figure 4.2 A model route to commercialisation: technology readiness levels v systems readiness levels. From James Baker ©2020.

Technologies typically start at the bottom left of the graph, low TRL and low SRL and then they slowly progress along the TRL axis until around TRL 5 or TRL6. This is then often the point that industry typically starts to engage and investigate on how they are going to take the technology into the various applications needed to access a market. This process is assumed to be linear and can be compared to a classic supply chain model—in this case, a "science supply chain" that is taking new knowledge from the lab ultimately into the marketplace. For example, this journey would begin in academic research groups based in a university like Manchester and then transfers into a specialist research institute and from there onto an innovation accelerator. What is important to recognise is that this science supply chain model will not necessarily facilitate a single flow of outputs. A project could hit a setback along the industry readiness journey and have to retrace its steps as the problem is unpicked and looked at again. (Similar to an unlucky counter having to slide backwards in a game of snakes-and-ladders and then having to start its climb all over again as described by Professor Phil Withers, Chief Scientist at the Henry Royce Institute, see Fig. 4.3). In reality, you can often appear to be progressing up the TRL levels to maybe four or five and then suddenly realise that what you have produced is not manufacturable, or it cannot be scaled up at the right cost, or does not meet a certain classification or certification and hence the need to go back to a lower TRL level in order to address some of these fundamental issues. This, in fact, is the strength of this science supply chain model because this linear, symbiotic relationships between all stakeholders, whether academic or more commercially focused, provides a shared interest in the final output.

The ups and downs of the 'science and innovation' supply chain

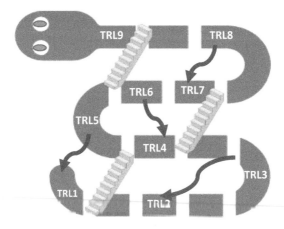

Figure 4.3 The science supply chain: the snakes and ladders-style progression along the technology readiness levels. From Prof. Phil Withers, Regius Professor of Materials at The University of Manchester and Chief Scientist at the Henry Royce Institute.

4.5 Creating the Manchester Model of Innovation

The whole process of advancing through the TRLs and SRLs from bottom left to top right in the curve is what typically can take many years and, indeed, as discussed earlier many decades. Even then, once you are at the top right of the curve for a product, to make any changes or adaptations to the new application could take many more years of development to re-enter the market again.

This whole, painstaking process would have to be reformed if we expect to reverse the reputation that the UK was great at discovery but poor at commercialisation. But to take breakthrough discoveries from the labs of our great UK research universities requires linking our science supply chain with the right industrial supply chain, so enabling the domestic production of our new products and capturing the value from market success.

This lab-to-market reformation is the story of graphene's commercialisation—a "Manchester Model of Innovation" that would demonstrate how we could bring a brand-new material to application by creating a unique R&D community in the place of discovery. If we could challenge the existing business model and reduce the risks and timescale of getting a new breakthrough material to first products, then we could expect to attract investment and look to create commercial and economic value. This investment could come from industry or from venture capitalists or new investors, such as small businesses (SMEs) who might be better placed to develop new disruptive products and benefit from new market opportunities.

So, a key step in converting the phenomenon of "discovered in Britain" into "made in Britain" was a new national research institute dedicated to applied scholarship in graphene— the £61 million National Graphene Institute (NGI) facility which not only brought together the scientists from across The University of Manchester but started the interaction and collaboration with industry and the wider supply chain. When I joined the University in 2013 the NGI had just started its construction and the facility's design and capability were being led one of the Nobel laureates Professor Kostya Novoselov. Kostya played a key role in creating a facility that was going to be a key enabler, in not only the development of graphene but the whole family of 2D materials in the future. The NGI was to act as an accelerator, to bring together academia with industry in a collaborative environment (see Chapter 5).

The NGI was a critical first step but The University of Manchester was also fortunate to receive funding for a second graphene facility, partly due to its relationship with football. Manchester is world famous for two football teams "the Reds Devils" and the "Sky Blues". Manchester United (the Reds) probably historically have had more success however in recent times Manchester City (the Sky Blues) has started to see some significant success following the acquisition by investors from the UAE and the creation of the Etihad stadium and sponsorship through Abu Dhabi. It was around 2014 that an opportunity arose to seek further funding to build a second graphene facility, an innovation accelerator that what was to become known as the Graphene Engineering Innovation Centre (GEIC) (see Chapter 6).

It was hearing about the proposal for what was to become the GEIC—pronounced "geek"—that first attracted me to talk to The University of Manchester about the idea of me picking up the graphene commercialisation leadership role. The NGI was still under construction and the funding had not yet been secured for the GEIC, but it was the prospect of the leading and shaping the new GEIC facility that really had the scope and potential to transform the way we could take a new material like graphene from the lab and put it into the marketplace that really excited me. The £60 million GEIC facility was eventually co-funded by the Abu Dhabi Energy Company (Masdar) together with UK government and European funding. It was a great opportunity to take something that had never been created before to deliver unprecedented commercialisation opportunities for an advanced material at scale. However, while the GEIC might be a unique facility there were already great examples in the UK to take inspiration, such as the High Value Manufacturing (HVM) catapults. The HVM catapults are government-funded institutions that had been created to translate some of the great activity from universities into industrial manufacturing capability.

In addition to the UK Catapults, there were other great R&D models around the world to draw on for inspiration, including the excellent Fraunhofer network in Germany and the mighty Defense Advanced Research Projects Agency (DARPA) in the USA. So we now had the opportunity in Manchester to create not only an iconic £60 million facility at the heart of The University of Manchester—on the University's soon-to-be regenerated North Campus—but also a new, placed-based business and innovation model that we went on to call the "Manchester Model of Innovation" (see Fig. 4.4).

In designing the GEIC, we were to put great emphasis on the various applications that graphene had the potential to create; whether this was in terms of new features and functions of existing products or in the creation of new disruptive products and applications. These range from composite's through to energy storage, through to coatings and membranes technologies, as well as inks sensors and formulations. At the heart of this new facility we would also look to ensure that we had both the academic grounding and understanding of the fundamental science together with the ability to do rapid scale up in a factory-like facility that we refer to as a "make or break" capability.

While the GEIC was designed to apply and manufacture 2D materials, the intent was always to create partnerships and collaborations with industry and the creation of a supporting supply chain. We wanted to partner with industries who produced graphene not just at the gram and kilogram level but also produce it at the tonnes or kilometre squared as required by industry— and so we developed methodologies for both bottom-up and top-down processes at scale. Also, we continued to partner with key organisations like the National Physical Laboratories (NPL), plus the specialist equipment suppliers, who together provided key capability not only in measurement but also on how we could move into quality control and process control of a high-quality and consistent 2D material (see Partnership Case Study in Chapter 8). The application labs would be even more critical in terms of doing that rapid scale up at the kilogram or metre squared level of activity.

Effectively, we developed a unique "design, make and validate business model" that meant we could engage with industry around a real business challenge. We could respond to the business need by going directly to our GEIC laboratories to rapidly produce, for example, a batch of samples of a material mixture or formulation, which could then be quickly measured and characterised to see if the material was validated in terms of the expected performance. If the material was acting as expected, we could then proceed into a rapid spiral development programme.

Alternatively, if the experiment failed, we could just as quickly understand why that experiment had failed. For example, if this was as a result of a fundamental technical reason then we could stop the work and try another process. Or if the development team had done something wrong then we could quickly learn and repeat the experiment in the correct way and move on. The ability to deliver this rapid "make or break" innovation model to support challenge-led research was to be a critical influence not only to the physical design of the GEIC building but also to the facility's whole operation model. And, as I was designing this pioneering accelerator, I was also aware that the GEIC was an importance piece in bigger strategic jigsaw that was looking to deliver innovation in a unique way for the UK.

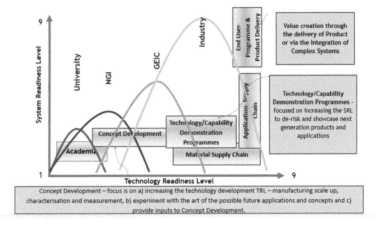

Figure 4.4 The Manchester Model of Innovation: the model route to commercialisation identified in Fig. 4.2 can now be fully realised thanks to the investments made by The University of Manchester and its partners to take graphene from the lab and into the marketplace. Scientific discovery is made by the academic community at The University of Manchester (mapping onto the red graph); research concepts are developed at the National Graphene Institute (NGI) (the purple graph); this applied research can then be scaled up and demonstrated as prototypes in the Graphene Engineering Centre (GEIC) (the green graph); and from this accelerator the new technology can be made fully market-ready by industry partners (the yellow graph). From James Baker © 2020.

Chapter 5

Building the First Home for Graphene: the NGI

By James Tallentire

A café in Didsbury, a suburb of south Manchester, was the innocuous spot where early but ambitious plans to commercialise graphene were hatched. As part of a Saturday morning routine, this pleasant coffee shop would be where Professor Luke Georghiou, now the Deputy President and Deputy Vice-Chancellor at The University of Manchester, would meet up with former Treasury civil servant Mike Emmerich, who was at this point Chief Executive of New Economy, Greater Manchester's own economic think-tank. Mike Emmerich is recognised as one of the key players in Manchester's pioneering devolution deals after being part of the Prime Minister's Policy Unit, where he was a senior advisor and held the brief on local government housing and planning.

From these informal discussions in Didsbury, the journey to commercialise graphene would include one of the most senior British politicians of the age, a brief encounter with a Catalonian soccer star and a business associate of Tony Blair. Luke said the driving ambition throughout these experiences was to find a way to accelerate novel materials to the marketplace. "Nobody has really successfully commercialised materials in terms of building an ecosystem so we are having to find ways to do it for advanced materials—there are, of course accelerators around but they are

---

Graphene: The Route to Commercialisation
James Baker and James Tallentire
Copyright © 2022 Jenny Stanford Publishing Pte. Ltd.
ISBN 978-981-4877-87-9 (Hardcover), 978-1-003-20027-7 (eBook)
www.jennystanford.com

mostly IT or tech-type things such as Tech Nation in London. And there are a number of accelerators in Greater Manchester but almost all of them are app-makers and people like that."

Luke's route to build this pioneering new model of materials commercialisation began with some blue sky thinking with his colleague and fellow strategist Mike Emmerich. "We knew each other and, as it happened, we used to casually meet in a café in Didsbury on Saturday mornings as both our routines meant we ended up there and we would start chatting about things. We got talking about the need to do something about graphene and came up with the idea that we'd have to go around the normal channels otherwise we just end up with single figure millions from the EPSRC [Engineering and Physical Science Research Council, UK's science funding body] or something and nothing else. So, we were talking, probably at that point, about a direct pitch to the then Prime Minister David Cameron; Mike, he came from the Treasury and very well connected there, so that was significant." Professor Dame Nancy Rothwell, FRS, the University's President and Vice-Chancellor, recalls: "A key point was a dinner hosted by (later Sir) Richard Leese, Leader of Manchester City Council, and (also later Sir) Howard Bernstein, then CEO at the city council, for Andre Geim and Kostya Novoselov shortly after they won the Nobel prize. It was at that meeting that we decided that Howard and I would write to George Osborne to seek funding for graphene commercialisation. Mike Emmerich and Luke had significant input into that letter which was quickly followed up with a positive response." Other people from the government's side who were involved in subsequent discussions were (now Lord) David Willetts, then the UK's science minister—who Nancy Rothwell had many discussions with—and also former Second Permanent Secretary to HM Treasury, (Sir) John Kingman. There were key involvements from senior University personnel such as Professor Colin Bailey, at that time Dean of the University's science and engineering faculty, who helped to lead on driving the implementation of the NGI and subsequent programmes; while his colleague Pauline Morgan, the former head of the faculty's finance team, supported with the investment part of the project.

Indeed, Alan Ferns, the University's former Associate Vice-President for External Relations and Reputation, also recalls that it was a serendipitous moment. George Osborne, as well as being

Chancellor of the Exchequer was MP for the North West constituency of Tatton; so it was in everyone's interests to make the Chancellor aware that a Northern city like Manchester was potentially home to an innovation that might rival any of those that the UK had brought to world in previous generations—DNA forensics, computing, the jet engine, atomic science, et cetera. George Osborne was part of the Coalition and succeeding Conservative governments that were to commit to a national industrial strategy and increased regional devolution, with the expectation of bringing new economic prosperity across the UK. Therefore, there was a meeting of minds over graphene, where regional and national agendas coincided to turn science into marketable UK innovation. "I can remember when the Conservatives came to power [in 2010 in coalition with the Liberal Democrats], they were looking for a high tech example to use in George Osborne's conference speech and we worked with their press office to include something in the Chancellor's speech about 'textile city' transformed into Graphene City, alongside a high profile visit to the University," explained Alan. Soon after, the government was to announce its intent to invest in graphene research, including a flagship facility in Manchester which was to be named the National Graphene Institute (NGI). George Osborne told the Guardian newspaper in December 2012 that graphene is "… exactly what our commitment to science and a proactive industrial strategy is all about—and we've beaten off strong global competition. Now I am glad to announce investment that will help take it from the British laboratory to the British factory floor. This shows that even in tough times we are investing in science which is vital to helping the UK get ahead in the global race" [1].

Osborne's rhetoric did turn to action with the investment of real money. However, added Luke Georghiou, the originally funding from the UK government was caveated, with some of the pledged money being separated off to go into a broader programme with EPRSC and only some of the cash coming directly to the National Graphene Institute project. "It wasn't going to be enough money for the building we wanted so we got the idea of going to the ERDF [European Research Development Fund] for the rest of it and they agreed. We had a visit from the European Commissioner. So that was how the NGI got going and it still has to report back to EPSRC talking about industry engagement."

With £38 million coming from government and the rest from Europe, building work began on the National Graphene Institute in March 2013 and officially opened by George Osborne two years later. Osborne proudly announced: "Four years ago, The University of Manchester came to me and asked the government to help make sure that Nobel prize-winning science conducted in Manchester leads to job creating innovation and discovery. Four years later it's fantastic to see that long-term vision become a reality with the opening of the National Graphene Institute. The new institute will bring together leading academics, scientists and business leaders to help develop the applications of tomorrow, putting the UK in pole position to lead the world in graphene technology" [2].

Plans for the new NGI building looked distinctive from the start. The new institute, costing £61 million, was a compact four-storey cube that occupies the full footprint of the corner site it sits on, so its front door opens directly onto the busy Booth Street East, which connects Upper Brook Street (A34) with Oxford Road. Its main cleanroom is located on the lower ground floor to achieve best vibration performance while a series of street-level windows allow curious passers-by to spy on graphene scientists at work. The building's offices and labs were to be intermixed on all floors, topped off with a roof terrace that the architects Jestico and Whiles hoped would form part of an al fresco social and public area. The building's inner skin was wrapped with a perforated stainless steel "veil" to provide a "unifying texture and fluid shape" made even more eye-catching as it is coated in dark, graphene-toned grey paint. The aim of this proposed building was to provide a home for the University's graphene research community—but not simply to pursue blue sky endeavours, there was a clear expectation that new knowledge in 2D materials would lead to practical and patentable applications by working with relevant businesses partners.

Indeed, the building would soon be hailed as an exemplar science building, especially as graphene pioneer Kostya Novoselov played a major role in the design to ensure the NGI went well beyond being fit for purpose. This was to be an experiment in designing a facility that would support great science but also be accessible to those from business communities. Architecture experts Albena Yaneva and Stelios Zavos, wrote a chapter dedicated to the NGI in the book Laboratory Lifestyles: The Construction of Scientific Fictions (MIT, 2019) [3] and they highlighted its unique ability to get the best out of those who use and visit the building, essentially creating a hothouse for new ideas and collaborative outputs.

"There is an amazing ontological symmetry that we witness: scientists, industry people, and 2D materials bear remarkable similarities to each other; they are all expected to form new bonds intensified by design. Thus, far from containing and sheltering scientific work, the building operates as a machine that maximizes the impact of graphene's implications and catalyzes the productivity of scientists and industry people, allowing for a faster exchange of ideas and creation of new ones. As graphene is evolving almost daily, there is a constant pressure for the building and its dwellers to adjust to ever-changing standards and the ecology of the graphene machine."

From their observations of the NGI in action, and conversation with its users, Yaneva and Zavos describe a science facility that has a work ethic, productivity and pace of delivery that would be the envy of any modern manufacturing company.

"The building is self-contained, argues Vladimir Falko, research director of NGI, and its strength is that it has many complementary facilities for the fabrication of devices (electronics and opto-electronics, nanocomposites, and printable electronic systems) that are all invented within its space, characterized and then developed as prototypes for possible industry use … Facilitating collaboration, the NGI speeds up technology transfer through the coexistence of labs and industry partners, and also accelerates progress."

After speaking to senior NGI scientist Dr Mark Bissett and the building's then operations manager John Whittaker, Yaneva and Zavos observe that: "The competitive aspect of their [graphene research community] work accelerates the rhythm of research and the labs are built to respond to this rhythm of development: they are flexible and could be used for different purposes; they allow easy connection to a gas line or to power through the gray spaces, and quick installation work. All this avoids wasting time and provides a smooth rhythm without disruptions to lab routines; as a result, the work tempo quickens and becomes more efficient. The in-house mechanical workshop avoids wasted time in subcontracting; the NGI technicians swiftly make changes on the spot."

Yaneva and Zavos conclude: "Our study of the National Graphene Institute in Manchester shows convincingly that modern laboratory buildings are vital settings for the active shaping of new patterns of research cultures, new socio-technical ecologies of innovation, and new alliances of science, society, and industry."

This new model approach to building a highly collaborative research environment is mentioned by Kostya himself during an interview as part of The University of Manchester's *Lockdown Lecture* series. He said about the burgeoning graphene community and its entrepreneurial approach: "There is technology transfer between different groups—recipes, samples, moving from group to another. There is a sense of unity in the community and it's really nice to see how people interact and it's even more enjoyable to see that really happening not because there is a structure but because people and students communicate on their own level. Just seeing new ideas developing from such a low level of communication is probably the most enjoyable and of course seeing new ideas being created that's always lovely" [4].

Without doubt the building has provided something different both in term of form and function. Another big admirer was Eli Harari, a graduate from The University of Manchester and founder of global technology company SanDisk. Luke Georghiou explained: "When Eli Harari first visited the NGI his comment was 'you have a really important engine here'. It was psychologically different. When you were bringing people into the University for graphene meetings they were originally held in the IT annexe for Computer Science and we were hosting meetings in rooms with no windows. We immediately got taken more seriously when the NGI was in place."

A science landmark had been created and the first phase of graphene commercialisation completed.

References

1. A. Jha. 'Super-material' graphene gets government backing, *The Guardian*, December 27, 2012. https://www.theguardian.

com/science/2012/dec/27/super-material-graphene-george-osborne#maincontent.
2. National Graphene Institute officially opens, HM Treasury press release, March 20 2015. https://www.gov.uk/government/news/national-graphene-institute-officially-opens.
3. S. Kaji-O'Grady , C. L. Smith, R. Hughes (editors). *Laboratory Lifestyles: The Construction of Scientific Fictions*, The MIT Press, 2019.
4. The University of Manchester's Lockdown Lecture featuring Kostya Novoselov, May 27, 2020. https://www.manchester.ac.uk/coronavirus-response/coronavirus-home-learning/lockdown-lectures/kostya-novoselov/.

Chapter 6

Steal with Pride: Creating the GEIC

By James Tallentire

The new and impressive National Graphene Institute (NGI) was now providing a beacon to all those interested in graphene. As Professor Luke Georghiou recalls: "Now all kinds of people started turning up and thinking they knew all about how to commercialise graphene, including some locally based industrialists who wanted to give us a flat amount of money to buy out the whole thing. But then most cooled off rapidly when they realised it didn't mean products in two years."

6.1 The Missing Piece

The NGI was able to anchor world-leading graphene research in Manchester and take it much closer to commercialisation—but it was becoming clear that the road from lab-to-market was missing something. Businesses that were now beginning to take an interest in graphene innovation were not committing to go all the way because they felt there was still too much of a risk for them. A big piece of the commercialisation jigsaw was missing; but at least the University knew what this piece should look like, said Luke. The Graphene Engineering Innovation Centre (GEIC)—as an acronym GEIC is deliberately pronounced "geek" as it was expected this

---

Graphene: The Route to Commercialisation
James Baker and James Tallentire
Copyright © 2022 Jenny Stanford Publishing Pte. Ltd.
ISBN 978-981-4877-87-9 (Hardcover), 978-1-003-20027-7 (eBook)
www.jennystanford.com

would attract attention in response to the word being admired in popular culture, as exemplified by tech entrepreneur Bill Gates and the hit TV show "The Bing Bang Theory"—would uniquely specialise in the rapid development and scale up of graphene and other 2D materials applications, while at the same time helping to de-risk this process. It was the missing jigsaw piece and its concept would be borrowed from other innovation catalysts, including America's legendary Defense Advanced Research Projects Agency (DARPA), the UK's Catapults programme and the Fraunhofer Society, the German research organisation with 72 institutes spread throughout the nation. This borrowing of influences has been dubbed "stealing with pride" by co-author James Baker (see Chapter 4), but in truth it was based on candid conversations and feedback from various leaders in the UK and international innovation communities. The GEIC was therefore designed to work in collaboration with industry partners and allow them to "create, test and optimise new concepts" in readiness for market entry. Risk would be mitigated, so long as lessons from failure could be quickly learnt. The proposition seemed certain to excite and attract would-be investors—but one problem remained, the £60 million cost of such a pioneering facility would match the price tag of the recently built NGI. Clearly this presented another huge funding challenge for the University in its ambition to deliver the next phase in graphene commercialisation.

6.2 Football Opens Up Leftfield Funding Opportunities

This impasse to bridge the innovation gap could have lingered indefinitely had it not been an opportunity from an unexpected source. The University's strategic relationships with the city council came into play again as this provided a connection to Abu Dhabi, the capital of the United Arab Emirates. This prosperous Gulf supercity is associated with the Premier League's Manchester City FC—and the club's majority owner, the Abu Dhabi United Group [1], has also branched out into other projects to support Manchester's wider community, for example, signing a 10-year partnership with Manchester City Council to revamp the east end of the city to build 6,000 affordable houses in the area as part of £1 billion deal [2].

"From leftfield came this idea—through Howard [Bernstein, former Chief Executive of Manchester City Council] again—that we could do something with Abu Dhabi," explained Luke. "Howard was at that time negotiating the big housing deal for East Manchester and he thought that they could engage with the University—at the same time they were doing a national tour with the UK DTI (Department of Trade and Industry), as it was called then, looking for things they could invest their sovereign wealth fund in. We had a couple of meetings with them at Manchester City football ground and the Lowry Hotel, I remember some of the football people were there, not the footballers themselves but people like Txiki [Begiristain, the former Barcelona and Spain player who is currently Director of Football at Manchester City]." Interest was piqued and meetings continued between the University and Abu Dhabi representatives—including former Tony Blair adviser Kevin Kokko, who went on to join Mubadala Investment Company, the Abu Dhabi-backed global investment fund [3].

From these conversations, the idea of the Graphene Engineering Innovation Centre, including its name, were hammered out. The result was a new partner for the University—Masdar, the Abu Dhabi-based renewable energy company owned by Mubadala, which is involved in projects across the world, including the London Array wind farm, one of the largest offshore wind farms in the world, and the Dudgeon Offshore Wind Farm off the coast of North Norfolk in East Anglia [4]. In landlocked Manchester, Masdar recognised it could be part of commercialising a game-changing new material and pledged £30 million to the creation of the GEIC. With this funding, a cornerstone was in place. Other investors followed, with £15 million from Research England, £5 million from Innovate UK, £5 million from European Regional Development Fund and £5 million from Greater Manchester Combined Authority. A total of £60 million, which meant the GEIC project was ready to go.

6.3   More Industrial than Academic

The GEIC was designed by world-renowned architect Rafael Viñoly, and when it opened in December 2018, it was the first new building on the University's North Campus for 40 years, signalling a new future for this area (more on this in Chapter 7) [5]. Design and consultancy firm Arcadis—which had also been involved in the creation of the National Graphene Institute and nearby hub building for the Henry Royce Institute, the national advanced materials institute—was appointed to provide project management and design expertise for the GEIC. They worked with Viñoly to develop a striking design that intended to allow the "public to witness the development of advanced materials through a highly glazed facade to showcase the research work being carried out within"[6]. The flexibility of the space, along with the extensive provision of essential services, process systems and future-proofing capacity within the core infrastructure design resulted in a truly adaptable space to support innovation with a range of industrial partners and R&D teams.

The facility, built by Laing O'Rourke Construction, was officially named the Masdar Building in honour of its main funder and its footprint covers around 8,400 square metres and features specialist labs to support key areas of applied research, including energy storage, composites, formulations and coatings, electronics and membranes. But the building's most distinctive feature is its high bay, where the facility can accommodate kit that is equivalent to pilot production processes. The 10 metre-high room is split into 9 metre square bays where the GEIC's commercial partners can install their large-scale equipment and have access to two 10 tonne overhead cranes for equipment unloading, installation and movement of materials—so the space looks far more industrial than academic.

The quality design of the GEIC was soon to be recognised when in 2019 the facility triumphed over almost 30 of the North West's most inspiring property projects to win not one but two accolades in the social impact awards organised by the global professional body Royal Institution of Chartered Surveyors (RICS), The prestigious RICS awards aim to showcase the most inspirational initiatives and developments in "land, real estate, construction and infrastructure". The GEIC was presented the "Design through Innovation" award before going on to win the overall "Project of the Year, North West" title, which recognised the scheme that demonstrated outstanding best practice and an "exemplary commitment to adding value to its local area"[7].

As Luke already mentioned, the University's proposition around graphene was taken more seriously when the NGI was opened. "But

I would say that stepped up again when the GEIC building opened because it's a statement that this is an industrial activity, you only have to go into the high bay to know you are not in an academic's lab. So, the buildings definitely project something as well as being functional and doing what they do."

No sooner had the paint dried, than prospective partners were knocking on the door of the Masdar Building clearly convinced that the GEIC offered the services and facilities to address their business challenges and fast-track 2D materials innovation. In fact, the GEIC signed its first three commercial partners even before the building had officially opened in late 2018. The Australian-based supplier of graphene products First Graphene; surface-funtionliased graphene specialists Haydale; and advanced engineering materials group Versarien were the first trio of partners—and many more were to follow (see Chapter 8).

This rush to join to GEIC has been impressive and it is recognised that these businesses are coming to GEIC because it is offering something different from the average university - or indeed the wider innovation community. For example, Dr Scott Steedman, Director of Standards at BSI and Editor-in-Chief of *Ingenia*, the magazine from the Royal Academy of Engineering, summed things up really well in an editorial published by *Ingenia* in September 2019 [8] when he said: "Six months after opening, the [GEIC] centre is already ahead of its plans to attract leading industry supporters.

"The main difference between the GEIC and Catapult Centres is that the GEIC is smaller and focused on a specific industrial outcome, the commercialisation of graphene, rather than covering a broad area, as the Catapults do in fields such as high value manufacturing, 'future cities', or the digital economy. This focus should enable the GEIC to build a portfolio of projects that develop their own momentum. As the resident engineering teams develop their expertise, so will their ability to respond rapidly to new ideas."

Scott Steedman then goes on to ask: "Could the GEIC become a model for other universities, strengthening the ecosystem in the UK to support new and emerging technologies?" And although, in response to his own question, he admits other universities like Sheffield, Warwick and Strathclyde have each established similarly successful ways to engage closely with industry, the Manchester model offers something different with its 'learn fast, fail fast' approach as it provides "... more flexibility for short-term, rapid turnaround projects, which makes it easier for companies to try out new ideas."

With the GEIC now operational, and working collaboratively with the NGI, a powerful campus-based innovation engine had now been launched which would be labelled Graphene@Manchester. As 2019 progressed both facilities were fast approaching their capacities, academic-led teams working in the NGI while business partners began occupy labs in the GEIC—and this revving engine was beginning to produce its first homegrown commercial success stories which combined academic expertise and a commercial ambition. One of the earliest milestones was the partnership with UK-based performance sportswear company inov-8, which worked with a team of composite specialists led by Dr Aravind Vijayaraghavan, Professor in Nanomaterials at The University of Manchester (see Innovation Case Study in Chapter 3). The pioneering Manchester Model of Innovation was working—but the story could not stop there. Already a new phase for graphene's commercialisation was being considered in a bid to take the programme beyond the campus and into a much wider ecosystem.

References

1. Manchester City FC Ownership. https://www.cityfootballgroup.com/our-business/ownership/.
2. City owner and council to build 6,000 new homes in £1bn deal. *Manchester Evening News,* 24 June 2014. Retrieved 6 October 2015. https://www.manchestereveningnews.co.uk/business/manchester-city-etihad-stadium-adug-7313788.
3. Tony Blair Adviser Kevin Kokko to Join Abu Dhabi's Mubadala Unit, *Bloomberg Business News,* December 17, 2014. https://www.bloomberg.com/news/articles/2014-12-17/tony-blair-adviser-kevin-kokko-to-join-abu-dhabis-mubadala-unit.
4. Masdar: Wind technology. https://masdar.ae/en/masdar-clean-energy/technologies/wind.
5. Masdar Building, Graphene Engineering Innovation Centre—The University of Manchester case study. https://vinoly.com/works/university-of-manchester-graphene-engineering-innovation-centre/.

6. GEIC case study on Arcadis web site. https://www.arcadis.com/en/united-kingdom/what-we-do/our-projects/uk/graphene-engineering-innovation-centre/.
7. Graphene Engineering Innovation Centre picks up two accolades in the RICS Awards: North West, press release published by The University of Manchester, May 2019. https://www.manchester.ac.uk/discover/news/graphene-engineering-innovation-centre-picks-up-two-accolades-in-the-rics-awards-north-west/.
8. S. Steedman. Lessons from graphene city, *Ingenia*, Issue 80, September 2019. https://www.ingenia.org.uk/Ingenia/Articles/367ca0c9-50f0-4202-ae4c-dcc1295eb59f.

Chapter 7

Graphene City and Beyond: Building an Innovation Ecosystem

By James Tallentire

With the completion of the National Graphene Institute (NGI) and the Graphene Engineering Innovation Centre (GEIC) on the University's own footprint, Luke Georghiou said phases 1 and 2 of Manchester's graphene commercialisation strategy were complete. The next stage would take activity onto a much more ambitious trajectory, away from the comfort zone of the campus environment.

7.1 Recognising the Size of the Prize

To help navigate this next phase, the University secured the advice from a renowned global consultancy specialist to determine the future for graphene commercialisation. Luke explained that this work provided: "… some useable information on the size of the prize; they modelled quite carefully the economic potential for Greater Manchester for graphene and came up with some very big numbers on GVA, on start-ups and on jobs and we have used those numbers quite a lot since that time." To reach full potential, a vehicle was now needed to support the scale-up of a new generation of

---

Graphene: The Route to Commercialisation
James Baker and James Tallentire
Copyright © 2022 Jenny Stanford Publishing Pte. Ltd.
ISBN 978-981-4877-87-9 (Hardcover), 978-1-003-20027-7 (eBook)
www.jennystanford.com

graphene-based businesses to run alongside the University-based infrastructure to translate and scale up innovation.

"It's dead easy to have a start-up, but should they can carry on with three people in some corner of the GEIC forever; not much point to companies like that. We want them to be able to scale up and turn into companies with big turnovers with lots of employees," said Luke. Luke's point echoed the seminal recommendations from entrepreneur and angel investor Sherry Coutu CBE who, in November 2014, authored the *Scale-Up Report* [1] which was commissioned by the then Coalition Government. Her report urged the UK government to support not only start-ups but also the scale-up of these businesses. "Competitive advantage doesn't go to the nations that focus on creating companies; it goes to the nations that focus on scaling them. One of the recommendations in my *Scale Up Report* is to boost the UK's investment ecosystem so these firms are not forced to look to the US or Asia for financing," explained Sherry Coutu [2].

To achieve this desired scale-up for start-ups in Greater Manchester, and the North of England more generally, an ambitious initiative has been launched, the Northern Gritstone Investment Company, with The University of Manchester teaming up with the universities of Leeds and Sheffield to work collectively to create a shared £500 million funding pot to support new enterprise. The aim is to create an offer that is similar in scale to those available to Oxford, UCL or Cambridge in a bid to deliver a step change in the North of England for the number and scale of viable spin-outs and IP income emerging from the region. Northern Gritstone has already set an ambitious target of creating three unicorns—privately held start-up companies valued at over $1 billion—within a decade. "A unicorn could take some time in terms of getting payback on a start-up, possibly even more than 10 years so it is a problem of patient capital," explained Luke.

Northern Gritstone publicly launched in April 2021 and has already attracted interest and support at the very highest level in UK government. In July 2021 the new company announced the appointment of former Goldman Sachs economist and Treasury minister Lord Jim O'Neill as Non-Executive Chairman, with investment trust specialist Duncan Johnson named as Chief Executive Officer. Lord O'Neil has been immersed in many matters related to the Northern Powerhouse project—the UK government-backed proposal to boost economic growth in the North of England—and is a champion for better recognition and investment into the region. Although Northern Gritstone will support all kinds of business ideas based on new intellectual property, the expectation is that graphene spin-outs will be among the first seeking major investment thanks to the acceleration provided by another new vehicle, the Manchester Graphene Company (MGC). Part of The University of Manchester's graphene ecosystem, MGC will effectively be a focused accelerator for graphene start-ups and spin-outs and be run by established entrepreneurial operatives that are steeped in business growth experience, such as commercialisation and investment expert Ray Gibbs; while the background intellectual property will be managed by the University's own IP agency, the University of Manchester Innovation Factory.

7.2 Developing a City Centre Innovation Hub

This innovation landscape—of which graphene commercialisation is an established landmark—is set to be greatly expanded under a hugely ambitious University-led scheme for the city-region and beyond. Part of this vision is a £1.5 billion innovation district—branded ID Manchester—which is proposed to offer the North's most realistic chance of creating a new, world class economic powerhouse founded on very competitive innovative technologies and a highly skilled workforce. In June 2021, The University of Manchester announced Bruntwood SciTech, the UK's leading property provider dedicated to the growth of the science and technology sector, as its preferred bidder to deliver the new innovation district and to help develop the site at the University's soon to-be vacated North Campus (the former UMIST campus, a legacy institution of The University of Manchester).

It should be noted that ID Manchester is already home to the "UK's major site for graphene commercialisation" as its footprint will include the Graphene Engineering Innovation Centre (GEIC), which will itself provide a base for the Manchester Graphene Company (MGC). Co-located with graphene commercialisation will be another pioneering hub for research and commercialisation, the world-class Manchester Institute of Biotechnology (itself only a stone's throw from the neighbouring Masdar Building, home to the GEIC) which focuses on industrial biotechnology and industry-facing biomanufacturing. With these already established facilities, the sheer scale of the fully developed ID Manchester project will help put the North on the map, especially with its ability to fully capitalise on connectivity offered by the arrival of High Speed 2 (HS2) and Northern Powerhouse Rail. It should be noted Piccadilly Station is a short walk from the site, so it will offer an ideal location for research intensive companies, as well as national research and research translation facilities. Independent estimates suggest that ID Manchester will bring thousands of new high quality jobs to the region with the ambition that the first businesses will be occupying "meanwhile" space on the site in early 2022, and the first purpose-built premises will be complete as early as 2024.

According to John Holden, Associate Vice-President for Major Special Projects at The University of Manchester, the development of ID Manchester will therefore deliver a significant step forward in the University's ability to create new manufacturing start-ups, jobs and growth, and deliver spill-over benefits to the city-region's economy. But ID Manchester is not the full panacea. "Its city centre location means that ID Manchester would not be appropriate for all innovation active companies, particularly large manufacturing firms that typically require premises that have big floor plates and a lower cost per square foot. Large manufacturers also typically want easy access to major transport networks—such as the main M6 and M62 motorways—and to be based in an established cluster near other manufacturers and firms in their supply chains. Internationally, university-driven city centre innovation districts often have manufacturing hinterlands where large manufacturing businesses can be based, for example Route 128 in Boston, USA." These "out-of-town" manufacturing clusters can then provide a complementary offer aimed at more diverse target market.

"The ambition of The University of Manchester is therefore to create a similar manufacturing innovation ecosystem in Greater Manchester and across the North of England," added John Holden. "As such, the University is working closely with its partners at the Greater Manchester Combined Authority to develop a proposal for a specialist site focused on driving innovation in the manufacturing sector based around the University's research, services and training strengths."

7.3 Creating an Out-of-Town Manufacture Hub

Greater Manchester was one of the first of eight trailblazer places to be invited by government to develop a local industrial strategy which it launched in June 2019 [3]. The plan identified the city-region's strength in advanced materials as a critical driver to economic prosperity, an assertion that was also evidenced in a science and innovation audit of Greater Manchester and parts of Cheshire to analyse regional strengths and to identify mechanisms to realise their economic potential [4]. This report coincided with the UK's Department of International Trade (DIT) awarding the region High Potential Opportunity (HPO) status reflecting the area's world-class capability to bring innovation to the lightweighting of materials—a capability that is of interest to most manufacturing sectors and can be achieved by adding graphene to a range of composites. Further confirmation of the role of graphene and advanced materials in driving economic growth was highlighted in the Greater Manchester Independent Prosperity Review [5], which was established to undertake a rigorous assessment of the current and future potential of Greater Manchester's economy. In fact, this report went on to aid the drafting of the Local Industry Strategy. The Independent Prosperity Review recognised the potential of a "Graphene City" which could be "... founded on the unique presence of world-leading science in advanced materials (including at the National Graphene Institute), engagement with business, and the creation of start-up companies" in the city of Manchester. The report furthermore identified the need to "systemise the pathway" through higher technology readiness levels (TRLs) which would be expedited by the University's Graphene Engineering Innovation Centre and "hence to turn discoveries to applications". This would be augmented by more training and skills development to create a supply chain of talent and "graphene scientists" with additional training in entrepreneurship.

7.4 Made in Manchester

The region's Local Industrial Strategy states that Greater Manchester continues to have a strong manufacturing base, employing over 110,000 people and generating £8 billion of economic output each year. Made up predominantly of small and medium-sized enterprises (SMEs), the city-region's manufacturers have specialisms in advanced materials, textiles (which has a strong concentration in north east Greater Manchester), chemicals, food and drink (with a strong cluster in Wigan), and is developing capabilities in industrial digitalisation. The wider North West region is a substantial manufacturing and advanced engineering cluster, with specialisms in aerospace and energy and clear potential to absorb graphene and advanced materials. So although some sectors have disappeared from Manchester, such as the making of computer hardware (see Chapter 10), other sectors are still manufacturing in the region, including textiles, machine-making and electronics, as well as component-makers contributing to larger supply chains which serve international sectors like aerospace or the auto industry. Historically, strategies to support business growth have been sector-specific, but what was now needed was a more holistic approach that would galvanise the city-region's manufacturing base around key drivers, which included widespread adoption of new technologies such as digitalisation and advanced materials. This would ultimately require new supporting infrastructure and bold strategic leadership to focus the regional economy on the big global challenges, such as climate change, to capitalise on new market opportunities and consumer demand for sustainable materials.

Concluding in 2019, the Greater Manchester Independent Prosperity Review (IPR) highlighted the conurbation's key strengths in advanced manufacturing, and that the manufacturing industry (and sub-sectors) present are highly productive: with no district or sub-sector under 80% of the UK average GVA per worker, and particularly in productive activity in Manchester, Salford and

Trafford. The IPR also points out the strength and speed of change in digital technologies and its diffusion across the economy in Greater Manchester, including the extent to which digitisation has been embedded within manufacturing, logistics supply chains and retail.

"What manufacturing is left in the UK is often either very productive or very embedded locally," explained Lisa Dale-Clough, who is Head of Industrial Strategy at the Greater Manchester Combined Authority (GMCA). "There have been so many challenges to the manufacturing sector over the past 30 years, so the modern UK manufacturing industry sector can be very robust. When you think about what elements of manufacturing have remained in the UK, you need to ask the question 'Why?' Quite often, there is a link to something technical such as design, that UK Plc adds value to, or they are part of a bigger, more strategic supply chain such as aerospace.

"For example, in Greater Manchester we still have a big textiles, clothing and materials manufacturing base in the city-region. And some parts of that sector have been modernised through the adoption of digital and e-commerce—and that's why you have seen the headquarters of [fashion brand] Boohoo, places like that, being located in this city-region. Other parts of that sector, for example around automotive textiles, have clear commercialisation opportunities for graphene. We already have lightweighting HPO status and we also have a big manufacturing base in electricals, components, machinery and semi-conductors."

7.5 Winning Hearts and Minds

"But although, through graphene, Greater Manchester has earned international recognition in advanced materials its local industrial relevance is not always clear," admits Lisa. "Sometimes I've had short shrift for having graphene front and centre in the Local Industry Strategy because people say 'how is that relevant to this small manufacturing company?' but when you take them through the potential applications of graphene and the machinery and technology needed to commercialise it, they see the relevance immediately. A name only gets you so far with an audience you have to go beyond that, to the potential applications to win the hearts and minds of the average manufacturing company."

7.6 Levelling up Graphene Opportunities

One of the key concepts in the Industry Strategy is about levelling up graphene's R&D opportunities, which originated and mainly based in the city of Manchester (i.e. "Graphene City") and sharing them in other parts of the city-region (i.e. creating what has been labelled by some as an "Advanced Materials City"). So, as described by University innovation strategist John Holden, future investment has been earmarked as part of the M62 Growth Corridor, an area that includes communities in Bury, Oldham and Rochdale which in political and economic terms could be at risk of being labelled 'left behind' communities. The ambition of the Local Industrial Strategy is that all this development across the city-region will be supported by technical expertise and investment from both national and local sources, with links to the university sector, local colleges and schools to complement and improve the existing skills base and help boost opportunities in Greater Manchester.

7.7 Materials for "Sustainable Manufacture"

The proposals outlined in the Local Industry Strategy for an "Advanced Materials City" have been added to by some strategic thinking that looks to respond to the regional call to Build Back Better in the wake of the Covid-19 pandemic and the national commitment from the current Conservative government led by Prime Minister Boris Johnson to "level up" R&D investment across all parts of the UK. Professor Richard Jones, Associate Vice-President for Innovation and Regional Economic Development at The University of Manchester, revealed some of this latest thinking as part of a GEIC partners' webinar streamed live by Graphene@Manchester in October, 2020. Richard described a new project called Innovation Greater Manchester (Innovation GM) which was officially launched in March 2021 as part of the city-region's bid to gain a share of the UK's multi-billion R&D investment as part of the levelling up agenda. Regional leaders are asking for £300 million to create Greater Manchester's own R&D agency, a proposal supported by businesses through the Local Enterprise Partnership, as well as having the backing of all five local universities. To work alongside ID Manchester, the regional project is looking to set up a manufacturing innovation park dedicated to materials, potentially on land off the M62 Growth Corridor (i.e. to support investment in communities like Rochdale or Bury—in fact, in July 2021, a consortium led by the UK metrology experts NPL, and including The University of Manchester, was awarded funding to establish in Rochdale the Advanced Machinery and Productivity Institute (AMPI) that will help transition the UK's machinery manufacturing sector in readiness for digital and autonomous systems). The theme for this manufacturing park reflects Manchester's national innovation leadership at the NGI and GEIC and will focus on "materials for sustainable manufacture". The park will boast high value manufacturing facilities and is expected to attract private sector labs as well as being underpinned by high value business services.

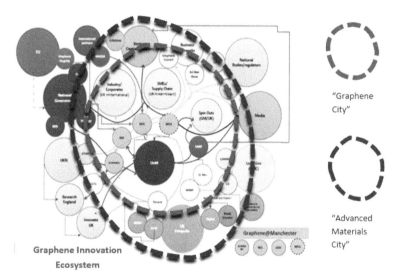

Figure 7.1 Visualisation of the Graphene Innovation Ecosystem: "Graphene City" and "Advanced Materials City" concepts sketched over. From James Tallentire.

Richard said it is expected that this new hub for "materials for sustainable manufacture" would essentially become a member of the High Value Catapult network and suggests that the Manchester park should focus on coatings, the materials science behind composites (as opposed to composite manufacturing expertise which is found at the Bristol HMV catapult), as well as textiles which (as Lisa Dale-Clough has explained earlier in this chapter) is still an important industrial sector for Greater Manchester, as well as neighbouring areas of Lancashire and West Yorkshire.

7.8 Regional Leadership in 2D Materials

To support the ambition to apply a Nobel prizewinning material to companies throughout the region, the Local Industrial Strategy established a specialist leadership group called the Graphene, Advanced Materials and Manufacturing Alliance (GAMMA). This new team is chaired by Jürgen Maier, the former Chief Executive of Siemens UK who is a leading UK industrialist, business commentator and national advisor. The Local Industrial Strategy says that GAMMA's role will be to identify opportunities to apply graphene and advanced materials technologies to address the UK's strategic industrial challenges. Therefore, there is an expectation to increase innovation, productivity and commercial growth by ensuring that the needed advanced technical and design skills would be available, along with the development of advanced manufacturing and materials sites across the city region conurbation. The new leadership group would also advice on inward investment and marketing, as well as help identify and mitigate the wider barriers to advanced materials commercialisation and manufacturing growth, such as access to finance or the availability of specialist premises and sites. This means GAMMA will have a defined spatial focus, as well as a focus on identifying market opportunities to commercialise graphene and advanced materials in Greater Manchester and the UK.

Lisa is convinced that the efforts of GAMMA and the Local Industrial Strategy will help underpin a globally competitive manufacture base across the Greater Manchester region, which in turn will contribute to UK Plc. She added: "Manufacturing is a strategic industry. It's how people think about their country — it helps build an international brand and it's how the rest of the world sees you."

References

1. S. Coutu. The Scale-up Report 2014, https://www.scaleupinstitute.org.uk/reports/the-scale-up-report-2014.
2. S. Coutu. On national importance of scale-up growth, quote reported at the Growth Investor Awards 2020 web site. https://growthinvestorawards.com/sherry-coutu-on-national-importance-of-scale-up-growth/.
3. Greater Manchester Local Industry Strategy. https://www.gov.uk/government/publications/local-industrial-strategy-greater-manchester-progress-statement.
4. Greater Manchester and Cheshire East Science: A Science and Innovation Audit Report. https://www.greatermanchester-ca.gov.uk/media/1136/science_audit_final.pdf.
5. Greater Manchester Independent Prosperity Review. https://www.greatermanchester-ca.gov.uk/what-we-do/economy/greater-manchester-independent-prosperity-review/.

Chapter 8

Making It Happen: Industry Engagement
By James Baker

The phased strategic approach to create a unique—indeed regional—innovation ecosystem around graphene, which has been described in Chapters 5, 6 and 7, is an impressive story of bold ambition, leadership and collaboration that has defined Manchester's achievements since the origins of the first Industrial Revolution. A critical element at the heart of that ecosystem—and therefore underpinning the whole Manchester Model of Innovation which was established around graphene commercialisation—has been the special relationship between the academic and industry communities. This bond is the glue that keeps everything together.

8.1 From an Academic-Led to Industry-Led Culture

The development of this academic-industry ecosystem is a critical part of what we have now defined as Graphene@Manchester, a campus-based community that provides the home for more than 350 academics and experts working on graphene and 2D materials, together with a total of £120 million worth of facilities and

---

Graphene: The Route to Commercialisation
James Baker and James Tallentire
Copyright © 2022 Jenny Stanford Publishing Pte. Ltd.
ISBN 978-981-4877-87-9 (Hardcover), 978-1-003-20027-7 (eBook)
www.jennystanford.com

equipment based within the National Graphene Institute (NGI) and the Graphene Engineering Innovation Centre (GEIC).

The National Graphene Institute played a critical part in helping to develop and broker partnerships with industry—indeed, the NGI was itself a "showcase" facility in its own right and helped to attract industrial interest and also prestigious VIP visitors to Manchester's graphene community. Among the high-profile visitors were China's dent Xi Jinping (see Fig. 8.1), on the final day of his state tour of the UK in October 2015, and then the following year, the NGI hosted HRH Prince William and Kate (the Duke and Duchess of Cambridge), who toured the graphene facilities in the early days of operation (see Fig. 8.2).

Figure 8.1 Xi Jinping, President of China, visits the National Graphene Institute (centre) in 2015 with from left to right former UK Chancellor George Osborne; Professor Dame Nancy Rothwell, Vice-Chancellor and President of The University of Manchester; Sir Andre Geim; and (far right) Sir Kostya Novoselov. Credit: The University of Manchester.

While the NGI was a key player in graphene's journey to commercialisation, the pace really started to step up once the Univerity had progressed with the completion of the GEIC facility in 2018. So, as well as supporting world-class applied science we started to work differently with industrial partners, originally around improving the process yields for the manufacture of graphene and then working to develop new products and applications. Because the GEIC operation was industry-led and offered a much higher TRL than you would expect in any typical university setting, our industrial partners said they felt very much at home. We also developed a very novel engagement methodology for working with industry; this "Catapult-like" model was designed to allow industry to sign up to a framework agreement that allowed partners to work on projects in a matter of days and weeks. This fast-track approach was something unprecedented in the way businesses traditionally deal with the university sector in terms of agility and timeframe. Using this framework, we signed a number of industrial partners in the first 18 months of operation, from companies who supply graphene and 2D materials through to end users, through to supply chain companies.

Figure 8.2 In the driving seat: Prince William visits National Graphene Institute (NGI) in 2016 and takes a seat in the graphene-enhanced BAC Mono car. Credit: The University of Manchester.

8.2 A Full House

The GEIC operates a tiered membership scheme to its business partners, with each level offering a varied menu of support services and facility access rights depending on the fee. During its first two years of operation the GEIC has successfully attracted interest and commitments from a range of sectors, including aerospace, steel-making, pharmaceutical, fashion and luxury, brewing, energy and energy storage, materials and textiles, packaging and printing, technology and electronics, coatings and infrastructure. We were hoping for one partner to sign up in readiness for our official opening in December 2018—in fact, we had three, **First Graphene**, **Haydale** and **Versarien/2D-Tech**, which became out first top tier members. These early partners either produced graphene material or functionalised this material for a particular product or application. Other companies to join as premium GEIC Tier 1 partners were to be more diversified and included **Gerdau**, the Brazilian-based steel company and one of the largest suppliers of steel in the world; and **GKN Aerospace Services**, the world's leading multi-technology tier one aerospace supplier. GEIC's Tier 2 partners now include **Airbus UK**, a wholly-owned subsidiary of Airbus which produces wings and other components for Airbus aircraft; **AstraZeneca**, the British-Swedish multinational pharmaceutical and biopharmaceutical company; **National Highways**, a UK government-owned company charged with operating, maintaining and improving England's motorways and major A-roads; **Vollebak**, a designer clothing company; **M&I Materials**, the Manchester-based materials company; **Qinetiq**, a British multinational defence technology company; **AEH Innovative Hydrogel**, a Manchester start-up company that is developing innovative hydrogels which will help shape the future of the agriculture and food industry; **Tunghsu Optoelectronic**, one of the earliest enterprises established to work with graphene in China; the global graphene network, **The Graphene Council**; **Nationwide Engineering**, a UK construction firm developing Concretene, a breakthrough graphene-enhanced concrete; **Nanoplexus**, a Manchester spin-out developing a platform technology (based around 2D material aerogels) for novel catalysts, composites and energy systems; and **Grafmarine**, a renewable energy business developing a new type of integrated solar power generation and storage system for the marine sector. Other partnerships have included **ARQ**, a producer of low-cost energy products; a large multinational brewer; the **Purico Group**, a major supplier of specialty papers to the world's food and industrial markets; **Gtechniq**, the automobile coatings and surface specialists; **Bullitt**, the UK based rugged mobile experts. The GEIC has also worked with a luxury Swiss watch manufacturer.

Despite having a number of partners it was critical that the GEIC business model remained that of being an "honest broker" in terms of graphene supply and did not offer graphene as a solution but to understand key requirements or challenges from our customers and then to match the most appropriate material to meet those needs and demands. This challenge-led approach was key towards the acceleration of graphene commercialisation and indeed it was independent of whether graphene might be the best solution. We could also work with carbon nanotubes, nano-platelets or even a non-carbon-based material that might, in fact, be most suitable—and this was all within the scope of what we looked to address as 2D advanced materials experts.

An example of how the GEIC—and Graphene@Manchester more widely—works with a large corporate partner is the international relationship with Gerdau, the giant Brazilian steel company. Gerdau is a major producer of long steel in the Americas and one of the world's largest suppliers of special steel, operating steel mills in 10 countries. When the company signed a Tier 1 agreement with GEIC, Gerdau CEO Gustavo Werneck heralded the deal as important in developing the company's "bureau of innovation in advanced materials".

The company's graphene projects have been led by Danilo da Silva Mariano, leader of Gerdau's Graphene and Advanced Materials R&D Centre. In a Graphene@Manchester blog published in July 2020 [1], Danilo explains: "We're looking at using graphene for anti-corrosion coatings, composites for the automotive industry, membranes and energy storage materials and technologies. We started to see a trend in the automotive industry, wanting to produce lighter cars—especially as a lot of companies are looking to make the switch from petrol to electric cars—and we knew we needed to reinvent ourselves in order to remain competitive.

"We looked all over the world for institutions to partner with and Manchester was mentioned by most of them as the ideal place to develop graphene applications. Nowhere else had the maturity of technology development and knowledge across such a wide breadth of subject areas.

"Sometimes it can take a long time to get contracts agreed and then never see any results, so we were really pleased with how quickly the University responded. Some of our projects will also transition from the GEIC into the National Graphene Institute—being able to work with two different facilities within the same university and having this structure really helps us to move forward."

Then Covid-19 made its impact in early 2020—Danilo added: "So, things might be slower, but they didn't stop, which is really positive. We're going to double the number of ongoing projects in the GEIC for the second semester and we just filed our first patent. The patent is so important for us. It was one of our first GEIC projects and was really promising."

Despite the global challenges, Danilo said he is really pleased about the "... rapid journey with the GEIC; arriving in September [2019], doing tests through to March and having a patent filed in the middle of the coronavirus pandemic. That's pretty cool!

"Gerdau is 120 years old as a company and this is only its 17th patent and the first that isn't for steel. So, this is a stepping stone for us—we're hoping to double the number of our patents in the next three years. Now we are consolidating all of this dreaming and innovation into products, so that's pretty important for in our history." In fact, so impressed with the R&D work being led in Manchester, the Brazil-based company launched a new enterprise called Gerdau Graphene in April 2021, which will develop and market products based on graphene applications. The new enterprise, which is now in the process of increasing its presence and capabilities in the GEIC, will work in partnership with the University as part of a global strategic alliance, with the aim of becoming a leading developer of graphene-enhanced products for a range of sectors, including construction, industrial and automotive lubricants, rubber, thermoplastics, coatings and sensors.

Another example of how GEIC model is the on-going work with National Highways, the UK government company that is responsible for the motorways and major A-roads in the country. This was one of GEIC's earliest of partnerships and the relationship began with a series of workshops that included the agency's supply chain businesses to help identify the business challenges and then to scope how graphene could potentially help find solutions. From this work, a series of exploratory work packages have now been commissioned to pioneer a range of potentially game-changing infrastructure solutions to improve deriver safety for millions of motorists (see Partnership Case Study 8.1).

PARTNERSHIP CASE STUDY 8.1
On the road to abolishing potholes and improving driver safety

National Highways has been working with the GEIC to see how graphene could help solve challenges that are faced by millions of motorists in their everyday lives.

Soon after the GEIC officially opened its doors the applications team started engagement with National Highways who were keen to demonstrate how innovation could be pulled through into the nation's road network to provide improvements and safety benefits.

National Highways is responsible for the motorways and major A roads in the country, which carry four million journeys over 4,300 miles of road. Initial dialogue was very positive with National Highways quickly agreeing to a Tier 2 partnership and for a short programme of activity, with the longer-term aim of developing a graphene-based bitumen.

This new road material would need to be more durable than existing products but also be elastic, so as to survive hot or cold expansion conditions which inevitably lead to surface cracks and ultimately potholes. This development would have a key benefit in terms of productivity for the road network –i.e. a reduction in potholes and therefore less repair and maintenance work.

National Highways were successful in bringing together their existing supply chain for a series of workshops where potential opportunities were identified and prioritised. These sandpit conversations have since evolved into a productive relationship with National Highways commissioning a number of projects.

The latest R&D programme, led by Dr Craig Dawson at the GEIC, will see how graphene applications can help improve the resilience of road surfaces; better protect infrastructure like safety barriers or roadside fences from the elements; help ensure road marking are longer-lasting; and even "blue sky thinking" around improved inclusion of electrical circuitry into our highways.

Paul Doney, Innovation Director at National Highways, said: "The opportunity to explore leading edge materials and what this might lead to for our road network is very exciting. The collaboration between Manchester and our operations teams, who understand the challenges out there on our roads, could lead to advances in safety and performance for our road users."

8.3   Bridging the Gap—Supporting Start-Up and SMEs

One of the great opportunities but also challenges of working with graphene and 2D materials is the variety of industrial partners we can potentially work with. We engage with both end-users and agencies like National Highways, through to original equipment manufacturers (OEMs) like Airbus, Rolls-Royce and GKN Aerospace but we also increasingly work with a whole series of small and medium enterprises (SMEs). These SMEs can vary from spin-outs coming out of The University of Manchester (and other institutions), through to small local companies, through to established companies of several hundred people—but still meeting the broad definition of an SME.

We were successful in winning funding from the European Regional Development Fund (ERDF) programme to support the development of SMEs based in the Greater Manchester area as part of a project we called Bridging the Gap. This was a key element in our strategy for engaging with small businesses and enabled us to understand how graphene might help existing enterprises create new products. Also, as well as supporting existing businesses we were also able to mentor a number of University-led spin-outs—but our focus was to see how we might also achieve rapid scale-up by using the GEIC as an accelerator. By looking at graphene as an additive, for example in a company's existing product range, there was opportunity to move very quickly from small scale into pilot production and into manufacture. See Partnership Case Study 8.2.

> **PARTNERSHIP CASE STUDY 8.2**
> **SpaceBlue: support for spin-outs and SMEs**
>
> One of the GEIC's early spin-out successes was SpaceBlue which was founded by Manchester academic and sustainability champion Dr Vivek Koncherry. Vivek launched his company with the aim to recycle waste tyres by converting them into new, high-value products after being enhanced with tiny amounts of graphene.
>
> "It all began when I first read newspaper reports that several thousand tonnes of waste UK tyres are being shipped abroad each year for disposal," explained Vivek, an expert in materials applications and new manufacturing techniques [2].

"I thought that needs to change and I became determined to find a much more sustainable way of using this end-of-life product."

SpaceBlue uses "chopped" waste tyres as its core feedstock and then by adding very small amounts of graphene—and some fresh rubber—you can produce a recycled material that is getting very close to the original performance of virgin rubber. Traditionally when companies have tried recycling tyres they have found that the product was "fragile" in terms of performance and tended to fray with small black rubber beads falling off the recycled product which, in itself, could cause potential harm and damage to the environment or user. SpaceBlue's clever use of graphene has mitigated that issue.

Within a year of joining the GEIC, Vivek's first product to hit the market has been the hexagon-shaped SpaceMat which can interlock to cover any desired floor area. The mats can potentially be used at the entrances of homes, offices, public and industrial buildings, as well as wider applications such as anti-fatigue or anti-slip coverings in areas like workplaces, gyms, playgrounds and swimming pools.

The project was a success of the EU-funded Bridging the Gap programme which enabled the GEIC to support a small start-up like SpaceBlue in its very earliest stages with advice not only on technical capabilities but also on marketing, PR and showcasing opportunities to successfully bring the new company to the attention of potential investors and customers.

"The innovation ecosystem at Manchester has been really supportive to someone like me who has a new business idea they want to take to market," says Vivek.

8.4 Setting a "Gold Standard" for Graphene

The Manchester Model, like any innovation accelerator is driven by a technology push but it was also increasingly important that it responded to real world challenges set by industry. If graphene was ever going to be commercialised, it was clear from the very beginning that we needed traction and interest from the supply chain and the companies who would manufacture and produce graphene material at scale. In these early years, graphene was being produced almost as a cottage industry and typically it was measured by the gram (or even milligram) or by the centimetre squared. We therefore needed to develop the methodologies for measurement, characterisation and for the creation of standards for this new material in readiness for its adoption at an industrial scale.

In this area we created our partnership with metrology experts the National Physical Laboratory (NPL) and we developed a best practice guide for not only what was graphene but also how you should measure it in terms of characterisation and performance. See Partnership Case Study 8.3.

PARTNERSHIP CASE STUDY 8.3
The "Greenwich Mean Time for Graphene":
Manchester becomes the global centre of graphene standards

When the National Graphene Institute (NGI) first opened, there were a number of suppliers of graphene material—but typically quantities were measured in grams or centimetres squared. The pace by which the scale of graphene manufacturing went from these small quantities soon escalated, moving on from using sticky tape and graphite which was the original manufacturing process. Rapid progress of graphene manufacture meant going from supplying by the gram towards the kilo and the tonne quantities.

A number of manufacturing processes were starting to be developed by partners to the NGI but also from across a global supply chain. Manchester's early work on standards proved to be critical as they would determine what "graphene was—and, even more, importantly what graphene wasn't". The term graphene is now essentially a shorthand for a variety of graphene forms and types. This includes graphene with different numbers of layers, different manufacturing processes and different properties that all give the material different features for different applications.

The University of Manchester and the National Physical Laboratory (NPL) formally joined forces to work together on the development of graphene standards, metrology and characterisation. As a result of their partnership the University and NPL published a good practice guide to tackle the ambiguity surrounding how to measure graphene—and this move has ensured the UK will become a leading authority on graphene standard. Effectively, this makes the NGI at Manchester the graphene equivalent to the Royal Observatory in Greenwich, a facility which has set the standard mean time from which all the world sets its clocks by.

The Manchester-NPL partnership also provides an independent service to characterise new graphene-related products, ensuring that product claims are supported by appropriate measurement and characterisation data, and that the claims made by vendors are reasonable and accurate.

With progress being achieved with standards, including agreements around nomenclature and recognised definitions of graphene, another important milestone was that our graphene supply chain started a collaboration under a REACH (Registration, Evaluation, Authorisation and Restriction of Chemicals) programme to ensure that any materials in development with us were compliant with EU chemical legislation, especially as we would be operating beyond the academic lab scale. Getting initial approval for REACH, was a key indicator of the progress being made in the graphene supply chain. So, despite there being many suppliers offering different forms and types of graphene, we were starting to see some improvements in terms of the quality of supply at a recognised standard—and at a more affordable price. As a result, we therefore made a critical step in graphene's commercialisation journey.

References

1. Beyond steel: How Gerdau is evolving through advanced materials, a Graphene@Manchester blog, published July 17, 2020. https://www.mub.eps.manchester.ac.uk/graphene/2020/07/beyond-steel-how-gerdau-is-evolving-through-advanced-materials/
2. Entrepreneur has sustainability challenge covered—with a SpaceMa; University of Manchester press release, December 19, 2019. https://www.manchester.ac.uk/discover/news/entrepreneurial-has-sustainability-challenge-covered--with-a-spacemat/.

Chapter 9

Getting Graphene Ready: Adopting the Manchester Model of Innovation

By James Baker

Nobel Laureate Andre Geim sometimes refers to the "brother and sisters" of graphene, which describes the wider family of 2D materials which now includes a plethora of single-layer materials and crystals. Following the isolation of graphene in 2004, researchers in Manchester and across the world have begun in earnest to study this ever-growing collective of 2D materials—and this academic work coincides with the development of a new supply chain to rapidly scale up the production of graphene material from a variety of different production methodologies and techniques.

9.1 Meet the Graphenes

A term that I have started to use to describe this extended family of 2D materials is "**Graphenes**". Not only are there many different types of graphene from single-layer to few layers to many layers of graphene, there is chemical vapour deposition (CVD) graphene and graphene nanoplatelets and also derivatives, including graphene oxide (GO) to name just a few. What we have seen, however, is that the science of graphene has now been applied to a whole range of other 2D materials—and the world of possibilities is almost endless

---

Graphene: The Route to Commercialisation
James Baker and James Tallentire
Copyright © 2022 Jenny Stanford Publishing Pte. Ltd.
ISBN 978-981-4877-87-9 (Hardcover), 978-1-003-20027-7 (eBook)
www.jennystanford.com

as we are now starting to see the creation of heterostructures, essentially a multifunctional stack made up layers of individual 2D materials placed on top of each other (see Chapter 11).

The growing family of 2D materials or Graphenes has now become so prolific they have now been broadly categorised (see Table. 9.1):

So how do we take what we have learnt from the Manchester Model of Innovation which has been built around the commercialisation of graphene and apply this across the whole sector of current and future 2D advanced materials? Well, a term that we have started to use in the Graphene@Manchester community is "getting industry graphene ready"—that is, we really need to challenge the way we have traditionally looked at the adoption of new materials in products and applications. The Manchester Model of Innovation is not just about the science of 2D materials but it is also about the development of a new translational system that can enable innovation cycles that measure in just days and weeks as opposed to the traditional innovation model for advanced materials that has typically been measured in months and years. If we can achieve this and reduce the risk of taking a new material from the lab to the marketplace, we therefore have the opportunity to drive productivity across a whole range of sectors, from lower manufacturing costs through to the enhanced performance of products and applications. Earlier we discussed the case study for aerospace (see Chapter 3) and how by applying graphene we can expect to use less materials content which would not only give a productivity benefit (i.e. lower manufacturing cost) but it can also offer product enhancements, such as lightweighting capability to reduce fuel consumption, noise emissions and other pollution.

Key to the Manchester Model of Innovation, however, is the focus on end application or "market pull". Graphenes therefore is a term that I believe can be used as a convenient shorthand for our new family of 2D advanced materials that can be pulled into a range of disruptive applications. Industry, to a large extent, does not care what material gives them a performance advantage, as long as it can be achieved competitively and repeatably at a cost and at a standard that is suitable for that particular market or application. Graphenes then represents a whole range of advanced or 2D materials that can have applications across a whole series of markets and products in the marketplace.

Table 9.1 Graphenes categorisation: the term Graphenes can be applied to the three themes or categories of 2D advanced materials

Categories	Description	Notes
1. Graphene and Graphene-Related Materials (GRMs)	Graphene is the original 2D layered material and at its most fundamental form is a single, one-atom thick layer. Other forms of graphene [1], defined as graphene-related materials (GRMs) by the Graphene Flagship [2] can be created by using various fabrication techniques. These variants include graphene films, graphene oxide (GO), reduced graphene oxide (rGO) and graphene nanoplates (GNPS).	Graphene and its various characterisations have now been officially standardised by the National Graphene Institute (NGI) and the National Physical Laboratory (NPL) [3].
2. GRAPHENE-LIKE MATERIALS The 2D Materials Family—founded on Graphene@Manchester research	Manchester's groundbreaking isolation and characterisation of graphene has inspired pioneering research teams to discover many other 2D materials from a range of elements. These can broadly categorised* as: • **Xenes:** a family of single-element graphene analogs that includes borophene, silicene, phosphorene, germanene, and stanine (the parent elements are clustered in the periodic table near carbon); • **MXenes** (pronounced "Maxenes"): a large set of metal carbides and examples includes the 2D, graphene-like sheets of Ti_3C_2;	

(Continued)

Table 9.1 (Continued)

Categories	Description	Notes
	• **Nitrides:** the most well know is **boron nitride** which is similar in geometry to its all-carbon analog graphene but has very different chemical and electronic properties; • **Organic materials:** using techniques in organic chemistry to convert multi-element organic molecules into 2D materials; • **Transition metal dichalcogenides:** made from a semiconductor sheets in which a transition metal atom is sandwiched between two chalcogenide atoms. *Categorisations taken from "2-D materials go beyond graphene", C&EN, 2017 [4].	
3. GRAPHENE-INSPIRED "DESIGNER MATERIALS" Hybrid heterostructures to create a new generation of advanced materials	*New generation of "designer materials"* that can now be created from a combination of any of the 2D layered materials described above in Categories (1) and (2). Hybrid properties can be leveraged by adding graphene to existing materials or by superimposing a mix of 2D materials into vertically stacked structures. The various layers are held together by the weak van der Waals force, creating so-called van der Waals heterostructures, which are ideal platforms to tailor novel materials with on-demand properties.	For more info about heterostructures, see Chapter 11.

Graphenes Categorisation: James Baker; James Tallentire ©2020.

9.2 Push or Pull

The pivotal relationship between "technology push" and "market pull" can also be compared to the relationship between technology readiness level (TRL) and sustainability research level (SRL) chart which I described earlier in Chapter 4 (see Fig. 9.1).

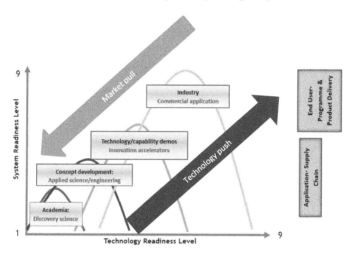

Figure 9.1 Technology push and marketing pull: the pivotal relationship between "technology push" and "market pull" can also be compared to the relationship between technology readiness level (TRL) and sustainability research level (SRL).

In the early days of graphene commercialisation, there was clear evidence of technology push activity which could be seen as the classic innovation scenario of a solution looking for a problem. To be honest, this stage of graphene commercialisation probably helped contribute to the "graphene is overhyped" comments then being made by some observers. However, what we have seen over the last few years is more of a "pull type" model which is thanks largely to an accelerator like the Graphene Engineering Innovation Centre (GEIC), which has been designed to specifically look at solving industry challenges or problems or a market

The focus of the GEIC is very much about maintaining our capability in graphene applications but also being prepared to produce 2D materials—Graphenes—if needed and perhaps not available in the current supply-chain. As we write this book, it is not only graphene that we are producing but one of our latest developments is to determine how we might scale up the production of MXenes (a large set of metal carbides), again this is something you will see and hear about more of in the future. So, with this broad application focus, we aim to solve industry challenges in a very different way than has been achieved in the past. We have developed our "make or break" approach where we undertake rapid design manufacture and validation and then look to iterate that development cycle or to fail fast and learn and move on quickly.

9.3 Plugging the Graphene Skills Gap

As well as a pioneering product developments, we are also looking to develop the skills and experience necessary for industry to manage the adoption of these future advanced materials applications.

9.4 PhDs in Graphene

An important talent supply line will be Manchester's own graphene focused Centre for Doctoral Training (CDT), branded Graphene NOWNANO. The nationally funded CDT programme aims to train engineers and scientists at the highest level with the specialist skills and knowledge needed to tackle the real world challenges faced by industry and society more generally. The Graphene NOWNANO course offers PhD students access to specialist and world-leading academics at The University of Manchester in partnership with Lancaster University; the four-year course will help this doctoral community drive research and innovation in the field of 2D nanomaterials. To help with this academic development (and also to build confidence), graphene PhD students are given the opportunity to present their own research to their peers at regular Friday afternoon seminars hosted by Nobel Laureate Andre Geim at the National Graphene Institute (NGI).

With access to world-class facilities like the NGI, PhD students enrolled on the Graphene NOWNANO programme will receive "state-of-the-art training" in the fundamentals of graphene and 2D materials, including potential applications and key techniques. This six-month training is followed by a three-and-a-half-year research project—and this academic commitment is complemented by direct engagement with a network of industrial and international academic partners. The course even has an enterprise component which was originally developed by innovation mentor Dr Simon Howell, Head of Innovation at the GEIC. Students are challenged to devise a hypothetical spin-out idea based around a graphene innovation. And a spin-out from this innovation element of the course was a project I am particularly proud to be associated with—the world's first Graphene Hackathon.

9.5 Innovation against the Clock

The Graphene Hackathon was held in November 2019 and organised by graphene PhD students. The hackathon attracted more than 60 students from a mix of disciplines and background who based themselves in the GEIC's Masdar Building, where they worked in teams of four to six persons continuously over a 24-hour period to develop ideas around a graphene ink application. Then, using some basic components and electronics, the teams were challenged to build a prototype device to test and showcase their innovative concepts. Nine out of the 10 teams were able to demonstrate a successful working prototype at the end of the 24-hour period, an impressive achievement. Participants were competing for cash prizes thanks to support from industry sponsors, including Google, IP specialists Mewburn Ellis and GEIC Tier 1 partner Versarien. To keep everything on track, experts from multiple disciplines were on hand to provide assistance and advice to the teams throughout the competition—and graphene pioneer Kostya Novoselov even dropped in about half way through the event to see first-hand how it was going and to add his encouragement to the frantic teams. The Graphene Hackathon was a very seminal experience, not only because it was the first event of its kind, but because it demonstrated that graphene-inspired innovation can be very accessible and should not have high barriers of entry. The very nature of the event, being run non-stop over a 24-hour period, really brought to life the "fail fast, learn fast" philosophy. It really was make or break as the teams were in a short race against time. Following the impact of Covid-19, the second Graphene Hackathon was held in April 2021 and was an all-digital affair; so rather than working in a single venue around-the-clock the teams were this time invited to take part in a virtual five-day programme of thought-leadership webinars, workshops and innovation challenges set by industrial sponsors and participants. I expect innovation events like this will help inspire a new generation of graphene entrepreneurs and help support a talent supply chain.

9.6 Enterprise Case Study: Hacking a Path to Innovation

Opinion piece authored by Bonnie Tsim, a PhD student at the Graphene NOWNANO Centre for Doctoral Training at The University of Manchester. Originally published in the September 2020 issue of Physics World *(IOP Publishing)* [5]

Academics are often derided for their lack of entrepreneurial skills. Partly that's because the curricula and training for undergraduate and postgraduate degrees are generally designed for those continuing in academia. Most students, however, prefer to explore other, non-academic career paths, in which the learning curve can be steep. Even students who do PhDs with an industry focus, for example at the UK's centres for doctoral training, will face unusual challenges.

With such limited exposure outside the academic bubble, what else can be done for students who are interested in what industry may have to offer? What opportunities are there for students to meet and talk with people from industry and other backgrounds? One exciting way to address such issues, I've discovered, is for institutions to run "hackathons". These were traditionally software-based challenges in which small teams of computer programmers and software developers created software. Those kinds of hackathons are still going strong, but they are now beginning to be used elsewhere with broader remits.

These hackathons usually run non-stop over a couple of days and can bring together people from a variety of backgrounds including science, arts, business and design to form interdisciplinary teams. In exposing researchers to an entrepreneurial, commercial and business environment in a very short space of time, they allow scientists to apply the skills they have developed during their degree in a different environment. Hackathons also highlight the importance of teamwork, resilience and communication skills.

Late last year [November 2019], a group of PhD students from The University of Manchester hosted the first Graphene Hackathon at the Graphene Engineering Innovation Centre. It was a 24-hour event where 10 interdisciplinary teams, each of between four and six people, designed, prototyped and pitched a commercial product using conductive graphene inks in front of a panel for the chance to win investment and cash prizes. The event required not only prototyping a technology that was worthy of investment but also developing business cases to outline the value of the products.

In developing the graphene-ink products, the teams faced some unexpected challenges including malfunctioning Raspberry Pis, screen-printing problems and flimsy final prototypes. During the event, industry experts such as strategy consultants, business development managers and patent attorneys were on hand to advise the teams as they worked throughout the night.

The hackathon was a fantastic way for participants to apply what they had learnt and put it into practice. I was involved with a team that created "BackUP"—an array of thin graphene strain sensors that can be printed onto a fabric seat cover for lorry, bus and truck drivers. It worked by monitoring the pressure on different parts of the seat in real time to indicate bad posture. If a driver was leaning on one side, for example, then it would send real-time feedback to remind them to improve their posture. The product, which came second in the competition, was designed to increase the comfort and wellbeing of drivers and reduce the number of days they are forced to take off work due to debilitating back pain.

9.7 Innovation and Skills

Graphene Hackathon case study—Part 2

Events like the Graphene Hackathon foster innovation by challenging participants in a competitive environment, boosting the likelihood of conceptualizing potentially disrupting technologies. For me, the experience highlighted the importance of a "fail-fast" approach that differs from the slower pace of academic research in which projects often run for months and even years. Fail-fast is often used as a mantra within start-ups as it highlights the importance of determining the long-term viability of a product or strategy. If something is predicted to fail, it's important to turn to a new idea without wasting more precious time and resources.

Indeed, the experience allowed me to apply the skills I developed from my physics degree to a commercial setting and made me realize the importance of thinking about fundamental research with a broader horizon. PhD students are in a unique place to develop their research and find commercial avenues for its applications. Hackathons can help widen the perspectives of participants, especially scientists who are looking to start spin out companies. The skills needed to build a successful business are not too dissimilar from the attributes needed to become a successful scientist as both stand on the foundations of perseverance, problem solving and presentation skills.

I believe hackathons could be run in many other areas of research such as energy harvesting from renewable resources, applications using recycled materials, and waste recovery. A hackathon that marries software and hardware is a particularly innovative way for early-career scientists who want to break out into industry-based roles to gain valuable first-hand experience in an emulated start-up environment. Here's to more hacks in the future.

9.8 Enterprise Award Helps Kick-Start Innovation

Another really successful mechanism supporting student engagement has included the Eli and Britt Harari Enterprise Award, which, in association with Nobel Laureate Andre Geim, is awarded annually to help kick-start commercially viable business proposals coming from Manchester's student community, including post-doctoral researchers and recent graduates. The award is co-funded by the North American Foundation for The University of Manchester through the support of one of the University's former physics students, Dr Eli Harari, founder of global flash-memory giant, SanDisk, and his wife, Britt. The competition recognises the role that high-level, flexible, early-stage financial support can play in the successful development of a business. The aim is to support the full commercialisation of a product or technology related to research in graphene and 2D materials—and to help achieve this, the award provides seed funding with prizes of £50,000 and £20,000 going to the individuals or teams who can best demonstrate how their technology can be applied to a viable commercial opportunity.

A recent runner-up was Manchester gradate Dr Beenish Siddique, who has founded AEH Innovative Hydrogel Ltd, a start-up company which aims to develop an eco-friendly hydrogel for farmers that would not only increase crop production but also has potential to grow crops in infertile land or as part of indoor "vertical farming" systems. Beenish has a passion for her business and its potential impact on agriculture, as she explains: "I believe there is an opportunity to change the future of farming. Globally, around 70% of the fresh water available to humans is used for agriculture and 60% of that is wasted; agriculture also contributes around 20% of global greenhouse-gas emissions. Our system helps control that waste and those emissions, shortens germination times and could enable an increase of 25% in crop yields."

After achieving success in the Eli and Britt Harari Enterprise Award in 2019, and being awarded a £20,000 cash prize, AEH Innovative Hydrogel Ltd went on to join the GEIC as a commercial partner where Beenish further developed her GelPonic hydrogel growth medium that looks to conserve water and filters out pathogens, while featuring a graphene sensor which allows remote monitoring. The potential of this pioneering technology has not gone unnoticed—and Beenish was awarded £1 million UK government funding through Innovate UK in the summer of 2020 to trial her graphene-enhanced growing system. Beenish's generous support from the Government is likely to be a response to the coronavirus pandemic, in tandem with looming net-zero targets, which have together sharpened the focus of policy-makers on investment in innovation. The Covid-19 pandemic has demonstrated the fragility of the UK supply chains, none more so than food supply, so indoor farming allows us to grow food in the UK that would normally come from another part of the world. That contributes to self-sustainability, reduces food miles and means we are not so reliant on international markets for our food.

As businesses—and even the government—look for solutions to the big challenges they are all facing I believe the ambitious commercialisation project at Manchester means we are now in a good position to support R&D communities to become graphene-ready. We have looked to extend our capability beyond graphene and into other 2D materials—the Graphenes—so we can offer a much more comprehensive capability; we want to be responsive to the needs of business and work with agility to deliver solutions while appreciating that new technology has to work within established systems. We also recognise that we have to nurture a supply chain of talent, producing enterprising "graphene graduates" who can join and support companies to manage the adoption of Graphenes or be part of the ecosystem by setting up their own businesses. This, I believe, is what makes the Manchester Model of Innovation unique and also very exciting.

References

1. H. Mason. Graphene 101: Forms, properties and applications, *CompositesWorld*, 10 January 2020. https://www.compositesworld.com/articles/graphene-101-forms-properties-and-applications.
2. J. E. Weis, A. Ahniya, K. Persson, A. Sugunan. Pre-study graphene characterisation and standardisation, 2016. https://siografen.se/app/uploads/2017/05/Graphene-Characterisation-and-standardisation-March-2017.pdf.
3. Graphene: Realising the potential of graphene and related 2D materials through our National Graphene Metrology Centre, NPL/National Graphene Metrology Centre. https://www.npl.co.uk/graphene.
4. M. Jacoby. 2-D materials go beyond graphene, *Chemical & Engineering News* (C&EN), 29 May 2017, 95(22). https://cen.acs.org/articles/95/i22/2-D-materials-beyond-graphene.html.
5. B. Tsim. Hacking a path to innovation, *Physics World*, September 2020 issue. https://physicsworld.com/a/hacking-a-path-to-innovation/.

Chapter 10

Creating an Icon: The Making of a Global Brand
By James Tallentire

Graphene's isolation in 2004 at The University of Manchester came at the right time in the right place. A perfect storm of academic curiosity, astute leadership and a favourable political climate all combined to ensure graphene ultimately became a global phenomenon; and within this mix is the University's own carefully crafted storytelling combined with an impressive commitment to commercialise this "wonder material" in the city where it was first isolated. All this, I believe, was greatly inspired by a desire not to have a repeat of the missed opportunities so often associated with British research and innovation.

10.1 The One That Got Away

The first time I really picked up on the lament surrounding an earlier Manchester-made technology that had "got away" was in an anecdote from Alan Ferns, the University's former Associate Vice-President for External Relations and Reputation. During a communications workshop held in November 2018 that aimed to refine the University's brand messaging on graphene, Alan made a presentation in which he mentioned how he had met, in his very

---

Graphene: The Route to Commercialisation
James Baker and James Tallentire
Copyright © 2022 Jenny Stanford Publishing Pte. Ltd.
ISBN 978-981-4877-87-9 (Hardcover), 978-1-003-20027-7 (eBook)
www.jennystanford.com

early career, Professor Tom Kilburn, the Manchester pioneer who led the development of the world's first modern computer in the 1940s Alan remarked that despite Manchester, and the UK more generally, being a post-war leader in an innovation that was to transform the world—holding onto that leadership was to prove more challenging. He said he hoped it would be different for graphene.

For me, the remarks were interesting, not least because Alan—who, in the mid-1980s, joined the precursor institution to The University of Manchester as a press officer—provided a link between two very different seminal innovations associated with the University. In fact, Alan was a key player in helping the University and the city to reclaim the origin story of modern computing, the world-changing technology which was invented in Manchester.

10.2 Birth of "The Baby"

The University of Manchester has a proud heritage of discovery in science and technology. Ernest Rutherford "split the atom" at Manchester in 1918 and inspired a revolution in nuclear physics and a generation later the University was host to a team of brilliant electronic engineers and mathematicians who would create the world's first stored-program electronic computer, which was officially named the Small-Scale Experimental Machine or more informally, "The Baby". This breakthrough machine was designed and built by a team of engineers that included pioneers F. C. Williams, Tom Kilburn and Geoff Tootill. The Baby made its first successful run of a program on 21 June, 1948, when it took 52 minutes to complete 3.5 million calculations before arriving at the correct answer—and in that process, The Baby became the first computer in the world to run a program electronically stored in its memory, rather than on paper tape or hardwired in. This was therefore the first machine that had all the components of a modern computer. From this Small-Scale Experimental Machine a full-sized machine was soon designed and built, the Manchester Mark 1, which by April 1949 was generally available for computation in scientific research in the University [1]. At the same time, the now famous World War 2 codebreaker Alan Turing had joined the University and became the Deputy Director of the University's Computing Machines Laboratory from 1948 to

1954. He helped with the programming of Manchester's prototype machines and from 1951 he worked in Coupland Building 1, which was a purpose-built annexe housing the new computing technology.

10.3   Losing Out to the USA

In the immediate aftermath of Manchester's early successes, Tom Kilburn and his own research team found themselves caught up in a computer "arms race" between the UK and the USA, who by then had set up innovation powerhouses like the Advanced Research Projects Agency (ARPA) [to be succeeded by the Defense Advanced Research Projects Agency (DARPA)] in a bid to gain global dominance in new strategic technologies. In the face of strong US rivalry, the Manchester group were to go on to create the UK's first "supercomputer", the Atlas, which was developed between 1956 and 1962. Two factors drove the Kilburn group, the first was an urgent need of the UK's nuclear physicists for more powerful computers because Britain had, by then, established the world's first civil nuclear programme and was a leader in this technology. The second was the realisation that the USA was planning its own new high-performance machines, such as the IBM STRETCH and the UNIVAC LARC projects. According to Simon Lavington, an expert on post-war computer history, it was feared that "… the early lead in computers that the UK had achieved in the period 1948–51 had slipped away" [2]. In an attempt to catch up with our American cousins, the UK government's National Research Development Corporation—a huge champion of the UK's early computer sector [3]—looked to encourage and sponsor British companies to rise to the challenge and in 1959 Ferranti, the Oldham-based electrical engineering and equipment company, joined the Kilburn project and the first production Atlas was inaugurated at The University of University on December 7, 1962, by Sir John Cockcroft, the Nobel prize-winning physicist who was then Director of the UK's Atomic Energy Authority. At the time of its introduction, Atlas was recognised as one of the world's most powerful computers and a total of six Atlas 1 and Atlas 2 computers were delivered between 1962 and 1966.

Manchester had given the world a technology that was one of the most transformational in the history of humanity. However, by the mid-1980s this story of how The Baby and its successors had made an historic contribution to modern day computing was fast becoming a faded memory while American behemoths like IMB and later Apple had eclipsed the UK sector. "F. C. Williams had died by then, although Tom Kilburn was still around. And to help tell their story was a very old plaque on the wall on Coupland Street saying 'this was where the computer was invented'. Occasionally, I took one or two journalists and television crews to view it; but we didn't make a big deal of it," recalls Alan Ferns, who was then a young press officer at the University. (Alan was to go on and climb the ranks at Manchester and went on to become Associate Vice-President for External Relations and Reputation.)

10.4 Reviving a Neglected Heritage

Alan recalls that this neglected heritage was soon to become vogue again. Personal computers were beginning to become ubiquitous in most peoples' lives, which begged the question—"where did these things come from?" Coincidentally, the secret work that was conducted at Bletchley Park, including the breaking of the Nazi's Enigma code which some believe helped shorten the war, was being made public and stirred the public's imagination. Another landmark came in 1986, when Simon Lavington published his seminal book "Early British Computers" which put the spotlight back on Manchester's contribution to computing, including the Atlas project.

To capitalise on this rekindled interest Alan teamed up with Manchester computer scientist Hilary Khan to plan an anniversary. Alan explained: "We took the opportunity in 1988 to do something on our 40th anniversary since the computer was really invented here. We worked with some people in the city but it was fairly low key—but we did involve Tom Kilburn in a couple of publications. We worked with our History of Science department to tell the story and there was quite a concentration of interest in 1988 when we started to take ownership of the invention of the computer. The city had some interest there and the story of Turing was also gaining wider public recognition. So, there was some interest at this time and that kind of grew over the following years. So, for 1998—which would be the 50th anniversary—Hilary and I decided we would do a big celebration and we began planning that, years in advance. We worked with the Computer Conversation Society and the city's Museum of Science and Industry."

Alan was able to reach out to many people who had worked on the original machines and also colleagues who had been part of the commercialisation of this exciting technology. One complication, recalled Alan, was the succession of corporate takeovers that befall the original maker of the Manchester computers. "The first machine had been commercialised by Ferranti, and Ferranti had been taken over by ICL—so when we were when planning the 50th anniversary celebrations we were dealing with ICL—but then ICL were effectively taken over by the Japanese company Fujitsu. So they were all involved."

The 50th anniversary event was certainly a much bigger extravaganza than the one from the previous decade. The focus of the University's celebrations was a so-called Launch Event, held at the Bridgewater Hall in Manchester's city centre and the programme included a short play to help tell the story about how those pioneers had built the "Manchester Baby" and a talk from Tom Kilburn who described the machine and the world-changing events on 21 June 1948 [4]. An exact replica of The Baby computer had been painstakingly built by enthusiasts in the city's Science and Industry Museum and in an exciting moment for those attending the anniversary event was when this facsimile was "switched on" via a remote satellite link from Bridgewater Hall. The replica machine can still be seen today by the general public at the city's science museum.

As a direct legacy of this anniversary activity, Alan said: "We took ownership of the story and really contested against the Americans and some other people who wanted to make a claim to have created the first computer. It gets difficult because you get involved with Babbage and what is a computer and what's not a computer—the world's first stored programmed computer is what we said we ran. We had to fight of interest from some people who were trying to build Turing's profile from saying he'd invented it; and actually, he didn't, he was the programmer of it. We were always firmly trying to locate it in electrical engineering, the guys who had actually built the machine, trying to give them credit; that is Williams and Kilburn. So, we did lots of work with them and it was very successful. Tom was an absolute delight to work with, very unassuming and a very, very nice guy. He would never had said this, but although he did return and keep in touch with people in the profession after his retirement, I don't think his achievement was ever properly acknowledged. Although coming quite late in life—he was in his late 70s at the time of the 50th anniversary—I think this kind of event reaffirmed his status and the importance of what he had done.

"But as part of all of that he [Tom Kilburn] talked about the journey from The Baby machine into manufacture of computers and there was a whole story to tell there about where the UK and Europe lost that journey from being in at the outset with Cathode Ray Tubes, computer storage systems, translating them into a commercial project, the Atlas machine, the Ferranti large scale machine, and somewhere along that trajectory it just got lost, the technology just got lost to America and Japan. A lot of other people who knew about these things were reflecting and what lessons could be learned from that."

10.5   A Manufacturing Retreat

For decades, the University and locally based manufacturers had been a very important part of the global development of computing. "Ferranti computing was based in Manchester, it became ICL and ICL had a big base in Manchester for a long time but, as I understand it, they carried on building big [mainframe] machines when other people had moved into desktop computers and micro-computers and processors and things like that," said Alan Ferns. "They [Ferranti/ICL] thought the market was 12 machines. Who would need these 'super brains'—the military, universities, perhaps the space programme. And when you look back at The Baby, the first Atlas machine—which all have a hundredth, maybe a thousandth of the power of my smart phone—there's another story there about what rewards or incentives flowed to Tom and those other pioneers to do that work; were they invited to the table to be involved in the commercialisation and the exploitation [of computer technology]. I think they were in the early stages but I think because they were engineers and academics, not necessarily further down the line."

Alan's final point suggests that research, innovation and commercialisation activities need to be integrated to achieve long lasting success.

How this bears out in the graphene story will probably help underpin the ability for this homegrown technology to bring indigenous benefits for generations to come. I can't help feel that those early computing pioneers would want their modern day counterparts working on graphene and its commercialisation to have a different experience. If we imagine our graphene innovators were to reconvene to celebrate the 50th anniversary of the isolation of the world's first 2D material in 2054 it would be satisfying to think that they could proudly boast how this new technology helped underpin an economic and manufacturing renaissance for Greater Manchester, the wider region and the UK as a whole. Lessons, it would seem, had been learnt from Manchester's computer story.

10.6 The Secret to Graphene's Universal Appeal

The graphene story—like the romance of Manchester's association with the innovators of early computing—has a resonance with mainstream audiences. The storytelling around graphene has been cleverly managed and greatly helped with assets such as two fascinating and engaging scientists.

10.7 The Quest for a Story with Universal Appeal

Graphene's ability to lend itself to universal storytelling was recognised very early on by The University of Manchester. Alan Ferns, a senior and influential communications and marketing leader at the University for many years, has been part of this narrative-building from the start. It all began, Alan recalls, when an appropriate research story was needed to mark the merger of The University of Manchester Institute of Science and Technology (UMIST) with neighbouring Victoria University of Manchester to create a new entity, The University of Manchester, which was officially launched on October 1, 2004. A new so-called "wonder material" was eventually selected as a flagship case study and splashed on the front page of the *Manchester Evening News* to help articulate how this new combined institution would be a northern research powerhouse. "We didn't know how significant or insignificant graphene was going be at that time; but interestingly, once we had done a scout around for technologies within the University we decided that graphene would be the one that would have some resonance with a broader public. We felt it would match their expectations of the kind of research breakthroughs the new university should be involved in." said Alan, who, from previous experience with Andre Geim, knew that this Russian-born, Dutch-British physicist had a dry sense of humour and was a good communicator. Andre already had a reputation for playful experiments and had levitated a live frog in 1997 to showcase his work in magnetism—and in turn helped him win the Ig Nobel in Physics in 2000, which is awarded to quixotic research that is seen as improbable but often inspires people to think more about the challenge under investigation. Geim also invented a new kind of sticky tape based on the adhesive feet of gecko lizards, which can walk up walls and hang upside down on ceilings [5].

10.8 The Game-Changer—Winning the Nobel Prize

What began as playful experimentation was to take on another dimension because around 2009 and into early 2010 Alan Ferns recalls that there was a rumble in academic communities that the isolation of graphene might be a contender for a Nobel Prize. However, many presumed Andre and Kostya only had an outside chance, because although fascinating, their work was perhaps a little premature to expect such an accolade. "To be honest, the University wasn't making great preparatory work or having big plans being put in place. The University had a great focus on recruiting Nobel Prize winners but I don't remember anyone saying we will soon be getting one or two of our own."

Then on the morning of the announcement for the 2010 Nobel for physics a series of telephone conversations was about to change everything. Alan received a call from University press officer Dan Cochlin (now Head of External Affairs at the Northern Powerhouse Partnership) to announce he had himself just received an unexpected call. "Dan got this call from Andre and he declares: 'I've just got a phone call saying we've won the Nobel Prize and they are going to announce it later today'."

Figure 10.1 A now iconic image of Andre Geim and Kostya Novoselov pictured together on the bench in the historic John Owens Courtyard at The University of Manchester. Credit: Russell Hart; The University of Manchester.

The seismic announcement galvanised the University's media operations. Alan had to pull Professor Dame Nancy Rothwell, President and Vice-President of The University of Manchester, from a high-level committee meeting to break the news to her and other senior leaders, including Professor Colin Bailey, who then headed the science and engineering faculty and already a big champion of graphene as a flagship research beacon. Frantic arrangements were made for a photo opportunity with Kostya still wearing his usual academic uniform of jeans and a t-shirt—and the resulting picture of the pair seated side-by-side on a bench situated in the picturesque quadrangle in front of the University's historic John Owen's Building has since become an iconic image. "Dan Cochlin did a very good job of handling the wave of interest and publicity around it all. We then had to plan for the ceremony and respond to all the international media working with the Nobel Committee," said Alan. "Very soon after the Nobel Prize stuff I said to Nancy [Rothwell] in PR terms, I think this is going to rumble on and become quite a big story." The rest, as they say, is history.

10.9 PR: Vital Ingredient in the Commercialisation Mix

As a communications professional with decades of experience, Alan admits he may be biased but nevertheless is convinced that by carefully managing the graphene narrative so early on—and making it a distinctive Manchester story—it greatly enhanced graphene's chances of going beyond being a great science story to one about innovation and commercialisation. He candidly says: "James Baker [CEO of Graphene@Manchester and co-author of this book] is an exception, I think he's brilliant with the communications aspect. However, I believe, in general, people in the commercialisation space miss the importance of the storytelling. But it's a critical factor in the ingredient."

Alan has no doubt on the role of compelling storytelling: "Building profile with professional communications has opened up some funding opportunities and partnership opportunities that just wouldn't have come. I don't think the politicians would have been interested unless telling the story in the right way; I don't think some of the investors would have been interested unless it was something they wanted to attach their name to. I think it played a big role.

"I believe you could have had the same technology, the same Nobel Prize, but if you had not curated or sought to enhance the narrative it could have been less impactful. Maybe I'm prejudiced; and maybe somebody else might have taken it and tried to make it their own. I think it would be interesting to know if this discovery had happened in Cambridge whether it might have gone in a different direction—for them; it might not have been so stand out and perhaps considered a more academic achievement and stayed in that space."

10.10 Managing the Story, Creating a Unique Vision

After that initial round robin of excited telephone exchanges between scientist Andre Geim, press officer Dan Chochlin and comms chief Alan Ferns, things would never be the same again. Dan would very quickly become the de facto media relations manager for this once niche area of the University's work in a bid to help manage the story and also safeguard Andre and Kosta from predatory journalists and would-be investors in the wake of their newfound fame. Alan explains: "Dan was very good at evolving his role as the agenda got broader, effectively he stopped being a press officer and became a brand manager, including public affairs and stakeholder liaison work. There are lots of examples at other universities where these things get away from the mothership—but because Dan was managed by me as part of our central resource we were able to make sure [graphene communications] kept closely connected and aligned with the University."

This early move to closely manage the graphene story would prove decisive. Eventually this work would develop into the Graphene@Manchester brand which was designed to represent and articulate the world-class output of the University's campus-based graphene community—both the pioneering academic work in 2D materials but also the commercialisation activities—to regional, national and global audiences. Like most brands, Graphene@Manchester has its own messaging to define its purpose and vision, with a top line that confidently proclaims:

> "It all began at The University of Manchester, which remains the world-leader in its application to transform society, create competitive advantage and deliver economic benefits from graphene and other 2D materials."

The next tier of messaging says: "Graphene and associated advanced materials will help drive significant and inclusive growth in the Greater Manchester economy, as well as create thousands of new, highly skilled jobs in the region." These statements demonstrate that The University of Manchester has a clear and public mission for its flagship work in advanced materials—that is, graphene is an economic driver to the city-region and the wider UK economy. The University is confident in its claims because it has built an impressive R&D community, with more than 300 researchers and support colleagues across more than a dozen academic disciplines collectively working on graphene and related 2D materials. More than £120 million of capital investment has created two flagship facilities for this graphene community—including the £61 million National Graphene Institute (NGI) and £60 million Graphene Engineering Innovation Centre (GEIC)—as well as a specialist teaching programme for graphene-focused PhD students called Graphene NOWNANO which has been established to create a talent supply chain to support this new graphene economy.

"Manchester is the home of graphene—and when you see the brilliant work and the products now being developed with the help of the Graphene@Manchester team it's clear why this city-region maintains global leadership in research and innovation around this fantastic advanced material," said Andy Burnham, the elected mayor for Greater Manchester and former Labour parliamentary front-bencher. Andy Burnham is an excellent example of high profile champion of Manchester's graphene story because essentially graphene is also Manchester's own story. In the way Detroit is the "Motor City", or Seattle is associated with aviation or California for new tech. Manchester is now a "graphene city".

10.11 Sharing Graphene's Alchemy: The Birth of the Beacons

As well as driving innovation in the world of industry, graphene has also helped pioneer a different approach to university communications. The new model calls for a disciplined focus on only a select handful of research themes or "beacons"—and this disruption to the established PR process was partly in response to a growing realisation that the graphene supernova was outshining everything else associated with The University of Manchester. This triggered a call to closely examine the graphene experience at Manchester to see if the same communications alchemy could be applied to other research disciplines. "People kept associating Manchester with graphene," said Alan Ferns. "Then in about 2012 we had a discussion at which the University's senior leadership team said "nobody knows anything about Manchester except graphene". Then we had a brave discussion in the University's senior leadership team on taking a closer look on why graphene took off and to see if we can't create some kind of "mini graphenes" and that's where the notion of the beacons campaign came from. So, we tried to take some of the ingredients of graphene and replicate that in some of other beacons in terms of resourcing and support but also developing a narrative," said Alan. With this thinking, the University embarked on an ambitious canonisation process to effectively reduce the hundreds of research areas into a mere handful of themes—a herculean challenge to a research-led institution like Manchester with a plethora of research groups and institutes all, understandably, seeking profile to raise recognition with peers, the public and potential funders for their own research areas. Taking lessons from the graphene success story, Alan said the final candidate areas needed to be able to be explained to the public, be genuinely groundbreaking and led by strong academic leaders who were willing to step into the limelight.

"Armed with these criteria, our University communications and marketing team began working with the senior leadership team to pick the areas," explained Alan, who added: "Now this is where the fun began. Everybody thought it was a good idea to pick a handful of areas, but nobody wanted to do the choosing and nobody wanted to miss out. To begin the process, we interviewed all the Deans and Directors of Research and came back with a list of 78 candidate areas. Some of these were duplicates or clearly non-starters, so with the help of the Vice-President for Research, we reduced this to a candidate list of 45 areas. This list was discussed at a senior leadership team workshop and we got it down to 19. At this point it became really difficult. We eventually got the list down to five areas because of the perseverance of Nancy Rothwell and because we persuaded colleagues that this was essentially a "marketing exercise" and not a measure of research quality or academic standing."

These final set of themes were "Advanced Materials" (with graphene the most high-profile flagship in this category), "Biotechnology", "Energy", "Cancer" and "Global Inequalities" (see Box 10.1 for more details on each). These five areas of investigation were selected because the University's leadership team was confident they featured a unique concentration of high-quality research activity. Also, crucially, for Manchester, a University that placed social responsibility at the core of its corporate vision, this suite of research beacons were to demonstrate a commitment to making a difference and "enabling us to improve the lives of people around the world through our research".

Box 10.1 The University of Manchester's five research beacons

Advanced Materials: graphene is very much the flagship in this category but the University has breadth of expertise in materials sciences, from biomaterials to materials for medical applications; from anti-corrosion expertise to developing materials for extreme environments, such as nuclear power stations, to antiviral materials that are of interest during the recent Covid-19 crisis. Manchester's reputation for materials science and engineering is so strong that it was the obvious choice to be the lead partner and host campus of the Henry Royce Institute, the UK's national body for research and innovation in advanced materials. This leadership in advanced materials is further underpinned by being recognised as a key economic driver for the city-region by the Greater Manchester's own Local Industrial Strategy.

Biotechnology: Manchester is leading the way, both nationally and across Europe, towards a bio-industrial revolution. The University is at the forefront of applied research into biotechnology, which looks to create the next-generation of chemicals, renewable energy and sustainable materials. Like graphene, biotechnology at Manchester is looking to scale up its commercial opportunities. The University is the lead academic partner for the Future Biomanufacturing Research Hub, which is backed by £10 million in government investment to develop new technologies to transform the manufacturing processes of chemicals, using plants, algae, fungi, marine life and micro-organisms—driving both clean economic growth while also making this technology more commercially viable. The investment is also part of the Local Industrial Strategy.

Energy: Manchester's expertise looks to boost the efficiency and viability of sustainable energy sources such as solar, wind, tidal and bioenergy; as well as guiding the UK's industrial strategy for the civil nuclear sector through the prestigious Dalton Nuclear Institute, the UK's most advanced academic nuclear research capability, which undertakes research across the entire nuclear fuel cycle, from innovative manufacturing techniques to waste management. The University is helping to ensure energy gets to the point of need efficiently, providing UK network partners with the knowledge to deliver reliable and sustainable power.

Cancer: Manchester's medical research ranges from understanding the molecular and cellular basis of cancer to the development and testing of novel drugs and other therapeutic approaches. Through nursing, psychology and policy work, solutions to the physical, emotional and economic impacts of cancer are being researched and put into practice across the University.

Global Inequalities: Manchester researchers are focusing on all aspects of inequality, from poverty to social justice, from living conditions to equality in the workplace. These experts seek to understand the causation of this inequality and to help mitigate its impact on communities across the world. Currently, Manchester academics are investigating how we can learn from the challenges of Covid-19, and build a more sustainable and equitable post-pandemic world. This in turn, partly links to the UK government's 'levelling up' agenda to better share wealth across the whole nation.

After senior leadership approved the beacons strategy it was up to the University's marketing team to deliver this ambitious project and in mid-2015 an operational project plan was drawn up which adopted the campaign strapline *"Global challenges, Manchester solutions"* which neatly reflected the fact the focus would very much be on promoting academic work that had real world applications. The phrase deliberately picked up on the priority grand challenges that have been identified by the UN or more locally by the UK government in ambitious interventionist policies like the Industry Strategy. The new plan also called for a consistent, results-orientated methodology that would identify relevant audiences, including business and innovation communities; government and policymakers; as well as the global academic community. "Phase one of the campaign was all about selling the concept internally and getting the research beacons quoted by our staff," Alan explained. "We were amazed how quickly and readily people took on board the concept. While phase two was about mapping some of the key audience groups for each beacon and establishing a small dedicated team to promote the beacons. This team consists of just six people, including a co-ordinator based in the central communications directorate and one communications professional for each beacon sitting closely with the relevant academic teams out in each faculty."

This new communications model was a significant change in direction for Manchester, moving away from what had been a comprehensive research communications strategy—i.e. drafting and distributing press releases in response to the publishing of new research in relevant academic journals—to a much more strategic and proactive way of working. Alan said: "I would say this new disciplined, campaign approach has transformed our research communications by allowing us to deliver more focused content for the beacons, including new online and offline collateral and content; dedicated social media assets; sustained editorial contributions to a range of University channels; and clear messaging for University presentations and exhibitions. Recent campaigns have involved everything from films, theatre productions, posters at Manchester Airport, exhibitions in San Francisco and a social media campaign aimed at international academics."

This activity was soon being recognised as sector leading, for example, the beacon campaign was awarded a Silver accolade in the 2017 Heist awards, the Oscars for the UK education marketing sector, in the category for the best marketing initiative to promote research excellence. The competition's judges said: "The University of Manchester demonstrated a strong approach to presenting a clear message in a research-led institution. This well-executed campaign supports a clear university direction."

Indeed, this clear direction was also, perhaps more importantly, being picked up by key influencers and their opinions are measured by the External Stakeholder Survey, an independently organised poll which has been completed for the University every two years since 2004. Those interviewed about the University's reputation includes policymakers in regional and national government; business leaders; gatekeepers to HE funding; senior figures in the health sector;

representatives from culture and media; as well as international agencies and partners. In the most recent set of surveys the beacons have been applauded as making the University's research portfolio more distinctive. Indeed, advanced materials and other beacons at Manchester are now seen as "the core areas of excellence" for the University. And if imitation is the best form of flattery then the Manchester model is indeed highly regarded. The universities of Glasgow, Nottingham, Loughborough and UWE Bristol are among those who have selected research themes which they describe as "beacons". Clearly, taking a risk on moving to a more focused communications strategy has paid off for Manchester says Alan, who adds: "We are still learning about what works and adapting our approach—but promoting our distinctiveness in this way is working for us."

10.12 Graphene, "An Icon of UK Innovation": Building a Global Reputation

Graphene's global market revenue is expected to top more than $400 million by 2026, claims a a research report by Global Market Insights [6]. According to the research the growing consumption of graphene in the electronics sector, especially in the developing economies, is helping to grow the international market. "Proliferating electronics and semiconductor industries in China, Japan, South Korea, the U.S. and India is one of the major factors boosting the overall graphene market growth," it is reported, and the article adds: "Furthermore, graphene is steadily penetrating in the automotive sector and gaining popularity in the production of light weight composite materials. For instance, in August 2019, [Liverpool-based] Briggs Automotive Company (BAC) launched a new lighter and higher-performance Mono R supercar which incorporated the use of graphene-enhanced carbon fiber in every car body component."

This huge potential growth in the graphene market in North America, Eastern Asia (China, Japan and South Korea) and Southern Asia (the Indian sub-continent) has informed Graphene@Manchester's international communications and engagement strategy (see Fig. 10.2).

Figure 10.2 Graphene goes global: international engagement has taken delegations from Manchester across Asia.

Graphene@Manchester is looking to leverage its international reputation as the "home of graphene" to optimise its presence in these fast-developing global markets—which also fits well with the internationalisation agenda of Greater Manchester. Lisa Dale-Clough, Head of Industrial Strategy at the Greater Manchester Combined Authority, talks about the asset of having graphene's worldwide reputation associated with the city. "We host a lot of international delegations and we tell a story of Greater Manchester which begins with the Industrial Revolution goes through Emmeline Pankhurst, to Alan Turing, the Co-operative movement, and always ends at graphene as the next phase of Greater Manchester's evolution—and whenever we tell that story its always resonates, the work that has been going on for the last 14 years or so has already had an impact internationally, which is great."

In recent years, the University has therefore worked very closely with MIDAS—Greater Manchester's inward investment agency—and, through this regional team, has also linked up with the UK government's Department of International Trade (DIT) which is eagerly promoting its "Global Britain" campaign and keen to showcase an "icon of UK innovation" like graphene. "Her Majesty's Government are champions of the great work The University of Manchester has been doing, especially the success of graphene," said Pukul Rana, Senior Strategic Communication Planner at Department for International Trade (DIT) in 2017.

10.13 Flying the Flag for UK Innovation

An example of this international partnership was when graphene was chosen to be the main theme of the award-winning UK Pavilion at Expo 2017 Astana, which was an international exposition hosted from June 10 to September 10, 2017 in Astana, the capital of Kazakhstan. This was first international festival to be hosted by a central Asian nation and the former Soviet Union republic is strategically positioned between Europe, northern and eastern Asia. The event exceeded expectations and attracted a reported 3.86 million people. Charles Hendry, the UK's specially appointed Commissioner for the expo, said that the role of the UK Pavilion was to celebrate Britain's great heritage and culture, which is recognised internationally—but also that the UK is still a global leader in innovation and he cited Manchester as an exemplar of this pioneering spirit and that graphene was the latest cutting edge technology from the UK. The pavilion showcase, which was designed by British architect Asif Khan and involved musician and artist Brian Eno, was visited by senior decision-makers from the UK and across the world, including Sir Alan Duncan, former Minister for Europe and the Americas; HRH The Duke of Gloucester; Ben van Beurden, the CEO of Royal Dutch Shell; senior Chinese government and business leaders; the President of the Swiss Central Bank; the Thai minister for energy; the Lord Mayor of London plus business delegates from the City of London. The Pavilion went on to win a silver medal for its beautiful exhibition design and exciting content. "This success owes much to the focus we put on graphene," added Charles Hendry. (The gold medal was picked up by Kazakhstan's Russian neighbour— although their pavilion did feature a real iceberg.)

Expo 2017 was a good example of how the UK government was able to provide platforms to promote Greater Manchester, the home of graphene, as a pioneer of a new technology with global potential. This national strategy was evidenced by the city-region being officially designated as offering high value opportunities (HPOs) to prospective investors across the world interested in the development of lightweight materials. This status was awarded in recognition of the region's expertise in both in graphene and light alloys, as well as its serendipitous location at the heart of the largest aerospace cluster in Europe and the second largest automotive cluster in the UK, offering easy access to end manufacturers such as Rolls Royce, Bentley, Airbus, BAE Systems, Vauxhall and Thales. The automotive, aerospace and rail manufactures are all driving the demand for lightweight—yet also strong and resilient—materials in a bid to reduce carbon emissions. The construction and energy sectors have similar demands. As a result the lightweighting opportunity is therefore expected to be worth £138 billion globally by 2021 [7].

10.14 Graphene in China and India

Manchester has had links with China for a number of years, which was helped by the profile given after the state visit of President Xi

Jinping to the UK, which included a tour of the National Graphene Institute (NGI) in 2015. This event was followed by President and Vice-Chancellor, Professor Dame Nancy Rothwell, with the support of graphene pioneer Kostya Novoselov, joining a high-profile British business delegation to China headed by then UK Prime Minister Theresa May. Also, region-specific trade visits have, for example, been led by Greater Manchester Mayor Andy Burnham who pitched graphene opportunities to business and civic leaders based in the city of Tianjin, a municipality and a coastal metropolis in northern China on the shore of the Bohai Sea. Graphene@Manchester has also looked to India and supported two successful government-backed campaigns to raise the profile around 2D materials. A delegation from Manchester, led by the Department for International Trade (DIT), travelled to India in October 2018 to promote the use of graphene ahead of the official opening of the Graphene Engineering Innovation Institute (GEIC) in the December. Representatives from MIDAS and Graphene@Manchester were part of the group to promote the opportunities to Indian corporates in Pune and Bangalore. Tim Newns, MIDAS's CEO, said the visit tied in with the regional economic strategy as Southern Asia is a key market for Greater Manchester and the graphene showcase was part of the "Make in India, Innovate in Manchester" campaign. The idea is to leverage the immense innovation strengths of Greater Manchester and the wider North to support Indian Prime Minister Narendra Modi's own "Make in India" economic policy. Graphene is the obvious flagship of Manchester innovation. After the success of the 2018 trip, a similar team re-visited India in 2019 to further build on the connections and opportunities made the previous year. The Manchester delegation met Indian businesses across a variety of sectors including aerospace, automotive and engineering, going back to Pune but this time putting Manesar and Hyderabad on the schedule. The delegation was again led by the Department for International Trade (DIT) and discussions were focused on applications in light alloys, batteries and technical textiles.

10.14.1 Graphene Goes to the USA

As well as raising profile in Asia, another important plank of Graphene@Manchester's international strategy has been to target the USA. Although some good work is being done by individual organisations—Ford, for example, are the first mass-market carmaker to use graphene in an everyday vehicle, the F150 pick-up truck, by adopting graphene-enhanced rubber components under the bonnet to help with soundproofing—it is generally felt that the USA has a less coordinated national strategy when compared with other parts of the world. As a result, America risked falling behind its peers. However, while America has arguably been late to graphene, there is still time to capitalise on the promise of these advanced materials—and the emerging markets are so big there is room to "cooperate, collaborate and benefit from progress in other countries" [8]. The special relationship between the UK and the USA is a good place to start, especially as both nations respect governance around trade and commerce. The global standards for graphene are evolving and there has been cooperation between the National Institute of Standards and Technology (NIST) in the USA and the National Physical Laboratory (NPL) in the UK. Encouraging connections between these institutes and the industries they serve will provide firm foundations for further Anglo-American co-operation on advanced materials. With these cross-Atlantic opportunities being nurtured, Graphene@Manchester felt it was time to make an official visit. In May, 2019, a Manchester-led graphene delegation found itself at the heart of the US policy community when it made presentations at the "Graphene on Capitol Hill" event as part of the wider American Graphene Summit hosted in Washington DC. The summit brought together US industry leaders and government agencies, plus key international figures—including this book's co-author James Baker representing Graphene@Manchester—to begin a "dialogue on shaping the global architecture surrounding graphene technologies and its impact on the US and global economy". To help capitalise on this policy event Manchester's graphene team decided to make a return trip just few months later in September 2019. James Baker and Graphene@Manchester colleagues joined forces again with trade and investment ambassadors from the DIT and MIDAS to follow in the wake of the jets streams of the RAF's Red Arrows as part of a high profile British delegation looking to make an impact with the American aeronautical sector (see Fig. 10.3). The Brits made their way to the Transatlantic Aerospace Symposium organised by the British-American Business Connections and hosted by Boeing to explain how graphene could transform this sector. Delegates were told how graphene's multi functionality could mean an aircraft wing of the future could be enhanced to make it lighter but also to provide lightning protection—so eliminating the need for separate copper-based lightning protection—saving further weight and complexity.

Figure 10.3 Top flight: James Baker, CEO of Graphene@Manchester (pictured left), and David Hilton, Head of Business Development (Advanced Manufacturing) at MIDAS, Manchester's inward investment agency, visiting the USA as part of a UK delegation visiting Seattle to attend the British American Business Connections Transatlantic Aerospace Symposium, which featured the Royal Air Force Aerobatic Team (aka the Red Arrows).

10.14.2 Graphene's Gravitas

Graphene has gone from a university research story to being adopted as a key driver to regional economic growth to being taunted internationally as the icon of UK innovation. Quite a story arc; but what underpins this brand-building achievement was the careful management of the story from the very start of its inception. This was to deliver PR and reputational advantages to The University of Manchester—but also giving graphene the gravitas that would impress investors, policy-makers and even world leaders. This in turn would help propel graphene from the lab along the commercialisation journey in the global marketplace.

References

1. How a 70-year-old 'Baby' changed the face of modern computing, UoM press release, 2018. https://www.manchester.ac.uk/discover/news/how-a-70-year-old-baby-changed-the-face-of-modern-computing/.
2. S. Lavington. The Atlas Story, 2012. http://curation.cs.manchester.ac.uk/atlas/elearn.cs.man.ac.uk/_atlas/docs/The%20Atlas%20story.pdf.
3. The National Research Development Corporation (NRDC) Collection. https://archiveshub.jisc.ac.uk/search/archives/36d1305b-cdcc-301a-bf93-d676ed108883
4. Manchester celebrates the 50th anniversary of the first stored-program computer. http://curation.cs.manchester.ac.uk/computer50/www.computer50.org/mark1/celebrations.html; http://curation.cs.manchester.ac.uk/computer50/www.computer50.org/index.html?man=true.
5. S. Connor. The graphene story: How Andrei Geim and Kostya Novoselov hit on a scientific breakthrough that changed the world ... by playing with sticky tape, *The Independent*, 18 March 2013. https://www.independent.co.uk/news/science/the-graphene-story-how-andrei-geim-and-kostya-novoselov-hit-on-a-scientific-breakthrough-that-8539743.html.
6. Graphene market value to hit $400 million by 2026: Global Market Insights, Inc., reported by PR Newswire, 27 November 2019. https://www.prnewswire.com/news-releases/graphene-market-value-to-hit-400-million-by-2026-global-market-insights-inc-300965977.html.
7. Lightweight materials—a £138 billion opportunity. https://www.investinmanchester.com/sectors/advanced-manufacturing/lightweighting.
8. In conversation with James Baker, Graphene@Manchester blog, 16 May 2019. https://www.mub.eps.manchester.ac.uk/graphene/2019/05/in-conversation-with-james-baker/.

Chapter 11

The New "Gold Rush": Graphene's Research Renaissance

By James Tallentire

While the commercialisation of graphene has been moving at pace, it should be remembered that the materials science that inspired this innovation has also been progressing and breaking new ground, promising yet more applications and technological advances. In fact, graphene's capability to continually excite researchers and offer new scientific breakthrough has even surprised its original pioneers Andre Geim and Kostya Novoselov. So, while the commercialisation of graphene has inspired the creation of a new innovation ecosystem, Manchester's pioneering work in 2D materials has similarly created a global science ecosystem that is constantly pushing the boundaries of blue sky and applied research in this rejuvenated field of science.

11.1 Living in the Graphene Age

In an interview Andre Geim gave to Adrian Nixon, Editor of the *Nixene Journal*, in January 2020 [1], the Nobel Prize winner looked to put the graphene and 2D materials phenomena into perspective. When asked why graphene had made such an impact on science and society, Andre replied: "Graphene is a new class of materials we were not even aware of just 15 years ago, it was completely hidden from materials science. And graphene is not alone, it has many brothers,

---

Graphene: The Route to Commercialisation
James Baker and James Tallentire
Copyright © 2022 Jenny Stanford Publishing Pte. Ltd.
ISBN 978-981-4877-87-9 (Hardcover), 978-1-003-20027-7 (eBook)
www.jennystanford.com

sisters, cousins, and by now we have probably studied dozens of those, hundreds of those materials.

"It's a new class of material and if we look at the history of the Human Race it's gradually built up from the Stone Age, to the Iron and Bronze ages, et cetera. We now live in the age of plastics and silicon, so I wouldn't be surprised that next we will be coming into the age of 2D Materials. But it is not that quick to transition from one age into another age because, as we know, one of the big problems for 2D materials is that we don't have tools—we are 3D humans so we have 3D tools, [therefore] it's very hard to deal with completely new types of materials. But [things are] gradually coming into line and industrial applications are gradually increasing.

"You know graphene has so many superlatives. You've heard about those—the strongest, the most conductive, the most this, the most that, and so on. But one superlative that is particularly relevant to this audience, I believe, is this material is the quickest we've jumped from an academic lab into consumer products. That is another superlative of this material."

11.2 The Graphene Gold Rush

The continued interest in graphene has been intriguing, even to the Manchester pioneers themselves. In the *Nixene Journal* interview, Andre says that in the "… first five years we knew practically everything about graphene itself, the material [had, we believed] been done and dusted by 2008, 2009."

However, around 2008, Andre recognised the trend for graphene to still attract large numbers of researchers and he noted: "With graphene, each year brings a new result, a new sub-area of research that opens up and sparks a gold rush" [2]. See Fig. 11.1.

This "gold rush" is evidenced by the global shifts in public research and innovation funding of at least $2.4 billion reported in 2014 and published in a case study by the Russell Group—the alliance of UK's leading research-based universities—entitled: *"Graphene: the transformative new material driving economic growth"*. The case study also references a 2011 study that found significant funding for graphene-related R&D in 26 European countries, while the USA, South Korea, Singapore and China have also committed substantial funds to graphene research and commercialisation [3] (see Fig. 11.1).

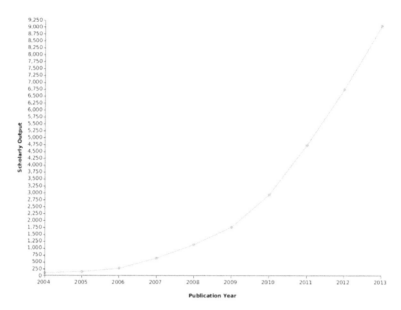

Figure 11.1 Graph showing exponential increase of academic articles published in the period 2004–13 from a search for "graphene" in the titles, abstracts or keywords. *Source*: SciVal and reproduced in *Research Trends*, September 2014 [2].

Kostya Novoselov looks at the graphene research phenomenon from more of a literary perspective. In the Lewis Carroll novel *Through the Looking Glass*, the young protagonist Alice is given a life lesson from the Red Queen who explains you need to keep running faster and faster just to keep on the same spot. To progress at all you need to accelerate even faster than you are running already. Kosyta made reference to this Carroll quote in his Nobel Prize biography and sums up how he still feels when working in the busy and highly competitive field of graphene research, an area he and Andre once had total knowledge.

11.3 Graphene's Not Dead

In the *Nixene* interview, Andre provides his explanation for this perennial interest and investment in graphene research—he says graphene has a capacity to periodically reinvent itself to offer something new to pique the scientific community. Andre said just as investigators felt research into graphene was "getting boring" and physicists and materials scientists, including those based at Manchester, were tempted to study other 2D materials he noted "astonishingly, graphene reincarnates itself every few years" to provide a whole new field of study. Since the original isolation of graphene there has been, explains Andre, a series of significant scientific milestones. "After graphene, there were graphene-like crystals and other 2D materials experiments; encapsulated graphene came along, offering the best system so far to study quantum physics in the electronic properties of materials. "Then, graphene set up a stage for what we call van der Waals heterostructures, including graphene superlattices, which perform very strongly without any sign of exhaustion. The latest reincarnation was two years ago with 'twisted' graphene that delivered carbon-based superconductivity, ferromagnetism and a myriad of other complex, many-body effects."

This chapter will look at some of these scientific milestones that have re-kindled interest in graphene research—as well as some external drivers and grand challenges.

11.4 Let's Do the Twist

Perhaps one of the most exciting milestones in graphene research has been twistronics, the name given to the study of how the angle—i.e. the "twist"—between layers of graphene or other 2D materials can change their electrical properties. In the interview with science journalist Adrian Nixon, Andre explained why this field was so seminal. Essentially, for him, the method has alchemical properties which could prove vital in the quest to find the Holy Grail of the electronics world—a superconductor that can operate at room temperature. The potential applications from this breakthrough could be transformational, including revolutionary electronic devices such as quantum computers, as well as new methods of data storage and improved energy transfer.

Andre said: "Twistronics is an interesting area of how flexible the system is in adjusting parameters because twist literally, rather than figuratively speaking, adds a lot of possibilities to change electronic structures. With a little bit of a twist [you] essentially go from one kind of metal to another kind of metal. [It's] like moving from say gold to uranium by a changing just by a fraction of a degree—the changes in electronic structure of this material is as dramatic as from lithium to uranium. This is very important to understand the fundamental properties of semiconductors and other phenomena [especially] if you say [this] has anything to do with reaching room temperature superconductivity."

This field of study, says Andre, offers huge potential and builds on his own observations of the impressive conductive nature of graphene which allows electrons to flow through it as easily as if it were a liquid—an observation known as "viscous conductance"—and potentially reaching limits physicists thought were fundamentally impossible [4]. Graphene as a "clean 2D structure" has far fewer imperfections than regular metals. As a result, electrons travelling through it scatter less and have the freedom to move faster and more easily—they effectively "shoot through the material, from one side to another".

The excitement around twistronics has inspired another research gold rush by providing the very latest graphene-inspired playground for physicists, with research teams such as the Jarillo-Herrero Group at the Massachusetts Institute of Technology (MIT) in the USA helping to identify the optimal twist as 1.1 degrees, a point that has been dubbed the "magic angle" [5]. After years of painstaking work the Jarillo-Herrero Group was first to demonstrate that magic angle theories were a reality—and immediately put new life into graphene research.

"Physicists are excited about magic-angle twisted bilayer graphene not because it's likely to be a practical superconductor but because they're convinced it can illuminate the mysterious properties of superconductivity itself. For one thing, the material seems to act suspiciously like a cuprate, a type of exotic ceramic in which superconductivity can occur at temperatures up to about 140 kelvin, or halfway between absolute zero and room temperature," wrote David H Freedman in *Quanta Magazine* (April 2019) [6].

A recent twist on the phenomena has come from an international research team led by Manchester which has revealed a material that mirrors the magic angle effect originally found in twisted bilayer graphene. The group led by Artem Mishchenko, Professor of Condensed Matter Physics at Manchester, has shown that the special topology of rhombohedral graphite—a form in which graphene layers stack slightly differently compared to the classic, stable hexagonal type of graphite—effectively provide an inbuilt "twist" and therefore offers an alternative medium to study potentially game-changing effects of superconductivity (2020) [7]. Interactions in twisted bilayer graphene are exceptionally sensitive to the twist angle, so tiny deviations of only about 0.1 degree from the exact magic angle will strongly supress interactions. Therefore, it is extremely difficult to make devices with the required accuracy and, crucially, that are sufficiently uniform to study consistently the exciting physics involved. The work of Mishchenko et al. (2020) on rhombohedral graphite has now opened an alternative route to accurately making superconductor devices. "It is an interesting alternative to highly popular studies of magic angle graphene," says Andre, a co-author of the study.

11.5   On the Wings of a Butterfly

After years of dedicated research, a group of scientists led by Andre Geim and Dr Alexey Berdyugin revealed an unorthodox phenomenon that was "radically different from textbook physics" and this work, published in November 2020 [8] has led to the discovery and characterisation of a new family of quasiparticles found in graphene-based materials. Called Brown-Zak fermions these extraordinary particles have the potential to create ultra-high frequency transistors which in turn could produce a new generation of superfast electronic devices. The Manchester study follows years of successive advances in graphene-boron nitride superlattices which has led to the observation of a fractal pattern known as the Hofstadter's butterfly—which describes the spectral properties of non-interacting two-dimensional electrons in a magnetic field in a lattice—and more recently researchers have reported another highly surprising behaviour of particles in such structures under applied magnetic field. "It is well known, that in a zero magnetic field, electrons move in straight trajectories and if you apply a magnetic field they start to bend and move in circles," explain Julien Barrier and Dr Piranavan Kumaravadivel, who carried out the experimental work. "In a graphene layer which has been aligned with the boron nitride, electrons also start to bend—but if you set the magnetic field at specific values, the electrons move in straight line trajectories again, as if there is no magnetic field anymore. Such behaviour is radically different from textbook physics."

Dr Alexey Berdyugin added. "We attribute this fascinating behaviour to the formation of novel quasiparticles at high magnetic field. Those quasiparticles have their own unique properties and exceptionally high mobility despite the extremely high magnetic field."

The authors propose Brown-Zak fermions—which were theorised decades ago—to be the family of quasiparticles existing in superlattices under high magnetic field; and the phenomena is characterised by a new quantum number that can directly be measured. Interestingly, working at lower temperatures allowed the researchers to lift the degeneracy with exchange interactions at ultra-low temperatures. Such findings are important for fundamental studies in electron transport and the Manchester group believe that understanding quasiparticles in novel superlattice devices under high magnetic fields can lead to the development of new electronic devices. The high mobility means that a transistor made from such a device could operate at higher frequencies, allowing a processor made out of this material to perform more calculations per unit of time, resulting in a faster computer. Applying a magnetic field would usually scale down the mobility and make such a device unusable for certain applications—but the high mobilities of Brown-Zak fermions at high magnetic fields open a new perspective for electronic devices operating under extreme conditions.

11.6   Graphene v the Graphenes

While graphene keeps making a comeback, Kostya says researchers have also been equally fascinated by the "brothers and sisters" of graphene, the fast-growing family of 2D materials (which co-author of this book James Baker has dubbed the "graphenes" originally in a *New Statesman* article [9] and referenced in Chapter 9).

INFORMATION BOX: 2D MATERIALS

Manchester's groundbreaking isolation and characterisation of graphene has inspired pioneering research teams to discover many other 2D materials from a range of elements. These can broadly categorised* as:

- **Xenes**: a family of single-element graphene analogs that includes borophene, silicene, phosphorene, germanene, and stanine (the parent elements are clustered in the periodic table near carbon)

- **MXenes** (pronounced "Maxenes"): a large set of metal carbides and examples includes the 2D, graphene-like sheets of Ti_3C_2

- **Nitrides**: the most well know is boron nitride which is similar in geometry to its all-carbon analog graphene but has very different chemical and electronic properties

- **Organic materials**: using techniques in organic chemistry to convert multi-element organic molecules into 2D materials

- **Transition metal dichalcogenides**: made from a semiconductor sheets in which a transition metal atom is sandwiched between two chalcogenide atoms

Source: "2-D materials go beyond graphene", *C&EN*, 2017 [10].

"We started to work on 2D materials very soon after starting work on graphene and for some time it was pushed a little aside as no-body had enough time to do everything," explained Kostya. "So, our first paper on 2D materials was in 2005 and we indicated that as well as graphene there were many other crystals as well—and since then the research picked up around 2010, 2011 when people started to work more on other 2D materials. Now it is kind of oscillating, so sometimes people come back to graphene and then sometime people focus on other materials, so it's very difficult to figure out what other trends [are developing]. Myself, I would say, I work 50–50 these days; 50 per cent on 2D materials and 50 per cent still on graphene."

Kostya is currently looking at the groundbreaking opportunities of stacking various 2D materials to make a nanoscale "club sandwich" called heterostructures. The concept actually goes back to the 1960s when semiconductor gallium arsenide was researched for making miniature lasers. Today, heterostructures are common and are used very broadly in semiconductor industry as a tool to design and control electronic and optical properties in devices. With the arrival of 2D materials a new generation of heterostructures have emerged in which atomically thin layers are held together by relatively weak van der Waals forces. The new, nanoscale van der Waals (vdW) heterostructures have opened up a huge potential to create numerous "designer-materials" and novel devices by stacking together any number of 2D layers. Hundreds of combinations become possible that were once inaccessible in traditional 3D materials, potentially giving access to new unexplored optoelectronic device functionality and other novel properties.

11.6.1 Making the Intelligent Club Sandwich

Kostya explains: "So, my current research interests are trying to create 2D heterostructures with pretty interesting properties, such as intelligent design. The intelligence works at different levels, so we use a little bit of AI to predict the properties of our materials—but also, we are trying to make our materials act intelligently so they can sit on sensors and actuators, with capabilities built inside the materials.

"We are creating complex materials called heterostructures when we combine several 2D crystals together to create an artificial material and we can very carefully find yielded properties because we can control the structure at the atomic level. We're working with all different layers of 2D materials and these days that includes dozens of different materials; there are semi-conductors, there are insulators, ferromagnets, super-conductors, you name it.

"You can make transistors with sensors placed together in one sandwich—you can do a lot of encoding of different properties at atomic level if you have access to one-atom thick materials."

11.7 Once Lost, Now Found: Magnetic 2D Materials

The growing range of 2D materials currently covers a vast range of properties but until recently this group had been missing one crucial member—2D magnets. However, this conspicuous absence has been rectified in the last few years with the introduction and study of a variety of atomically thin magnetic crystals; and this fledging field currently greatly excites Kostya.

Monolayer or few-layer materials with intrinsic magnetic properties were initially demonstrated experimentally in 2D ferromagnets, such as chromium germanium telluride ($Cr_2Ge_2Te_6$) and chromium triiodide (CrI_3) in 2017 which then opened the field to many other 2D magnet materials (2DMMs) being discovered and studied. Although the investigation of 2DMMs is still at an early stage, Ningrum et al. (2020) in a review of recent advances in 2D materials in *Research* (2020) predict that that they "… are promising building blocks for next-generation information devices, such as nanoscale spintronics and quantum technologies" [11]. While a paper review, co-authored with Kostya, makes a similar optimistic conclusion: "At this stage, the key questions that are being addressed are of fundamental nature, but as soon as 2D magnetic materials can be reliably synthesized with sufficiently high critical temperatures, the potential for technological impact is enormous." [12]. Interestingly, Sharpe et al. (2019) have also found evidence of ferromagnetism in twisted bilayer graphene layers. This surprising finding has revealed that magnetism can not only be manipulated but it can also be generated using van der Waals heterostructures made up of nonmagnetic 2D materials [13].

11.8 Rising Star in the Graphenes Family: Graphene Oxide

Graphene oxide, an oxidised form of graphene that is laced with oxygen-containing groups, has been dubbed the "raising star" of 2D materials [14]. "Graphene oxide (GO) is a graphene-based material that has gained significant interest in the last two decades due to its straightforward, scalable, and low-cost synthesis. It has been proposed for numerous applications: for instance, GO is a promising material as an efficient sieve for water remediation and for sustainable energy production via fuel cells," write Mouhat et al. (2020) [14]. Indeed, graphene oxide membranes have been found to be capable of forming a perfect barrier when dealing with liquids and gasses—and they can effectively separate organic solvent from water and remove water from a gas mixture to an exceptional level. To great media and public interest, a Manchester-based group led by Professor Rahul Nair has further developed graphene membranes and found a strategy to avoid the swelling of the membrane when exposed to water. The pore size in the membrane can now be precisely controlled or "tuned" which means they can sieve common salts out of seawater and make it safe to drink. The potential of this research has far reaching potential to support a growing global population as it faces water scarcity due to climatic changes and other challenges [15].

Another application of graphene oxide has helped Manchester researchers bring advances to wireless technology and the associated Internet of Things (IoT), as first reported in *Scientific Reports* (2018) [16]. The researchers have discovered that by layering graphene-oxide on graphene they can create a flexible heterostructure that can be developed into humidity sensors that in turn can connect to any wireless network. These graphene humidity sensors can potentially be mass-printed one layer at a time or at scale so making it very economical for manufacture. Dr Zhirun Hu, who led the work, said: "The excitement does not end with this new application here but leads to the future possibilities of integrations of this technique with other 2D materials to open up a new horizon of wireless sensing applications." While Kostya, who coordinated the project, believed this was the first example of printable technology where several 2D materials come together to create a functional device immediately suitable for industrial applications. "The Internet of Things is the fast-growing segment of technology, and I'm sure that 2D materials will play an important role there" [17].

11.9   Going for a Spin

The revolutionary capabilities of van der Waals heterostructure have also helped push forward the development of another exciting new field in the 2D world—spintronics, a radically different concept in electronics in which the rotational motion, i.e. "spin" of the electron (and its associated magnetic characteristics) can be used to carry or store information in a material in addition to its intrinsic charge. Andre Geim, a pioneer in spintronics research, summed up the phenomena: "The Holy Grail of spintronics is the conversion of electricity into magnetism or vice versa." When applied at nanoscale this exciting area of research has the potential to lead to the next generation of high-speed electronics and superconductivity. Spintronics big attraction is its ability to create low-power electronic devices that are not based on a charge current but on a current of electron spins. Spintronic devices could even enable electronics to go beyond Moore's law—the prediction made by American engineer Gordon Moore that the number of transistors per chip would double about every two years—by offering higher energy efficiency and lower dissipation when compared to conventional electronics, which relies on charge currents. So, in principle, the world could have phones and tablets operating with spin-based transistors and memories.

However, these transformational quantum capabilities of graphene are best achieved by creating an appropriate heterostructure. "As with a pizza, graphene technology is not only dependent on the graphene pizza base but also on its toppings," observes Bernhard C Bayer from the Institute of Materials Chemistry at the TU Wien in Vienna. "How these toppings are applied to the graphene is, however, crucial." [18].

A good example of getting you toppings right has been demonstrated by physicists from the University of Groningen who have constructed a two-dimensional spin transistor in which spin currents were generated by passing an electric current through graphene. A monolayer of a transition metal dichalcogenide (TMD) was placed on top of graphene to induce charge-to-spin conversion in the graphene. This experimental observation was described in the journal *Nano Letters* and *Science News* and *PhysOrg* [19] reported in September 2019: "Graphene, a 2D form of carbon, is an excellent spin transporter. However, in order to create or manipulate spins, interaction of its electrons with the atomic nuclei is needed: spin-orbit coupling. This interaction is very weak in carbon, making it difficult to generate or manipulate spin currents in graphene. However, it has been shown that spin-orbit coupling in graphene will increase when a monolayer of a material with heavier atoms (such as a TMD) is placed on top, creating a van der Waals heterostructure."

Dr Ivan Vera Marun, Lecturer in Condensed Matter Physics at The University of Manchester, has commented on these recent breakthroughs: "The continuous progress in graphene spintronics, and more broadly in 2D heterostructures, has resulted in the efficient creation, transport, and detection of spin information using effects previously inaccessible to graphene alone. As efforts on both the fundamental and technological aspects continue, we believe that ballistic spin transport will be realised in 2D heterostructures, even at room temperature. Such transport would enable practical use of the quantum mechanical properties of electron wave functions, bringing spins in 2D materials to the service of future quantum computation approaches" [20].

While Kevin Garello, a senior R&D scientist at Imec, and Stephan Roche, a research professor at the Catalan Institute of Nanoscience and Nanotechnology (ICN2), have mapped the potential commercial sectors where spintronics could make a big impact. They report in the Graphene Flagship's annual report for 2019 [21]: "Spintronics fosters the emergence of new markets across a broad range of fields. Aside from magnetic RAM technology, spintronics could play a crucial role in the development of new technology for motion sensing, mechanical engineering, computer games, robotics, fuel sensors, speed control, navigation systems and even minimally invasive surgery. Efficient spintronic devices have huge advantages over traditional electronic components: they have low volatility, low power consumption and density, and high data transfer speed. All of these properties are essential for the Internet of Things."

11.10　The Speed of Light

The disruptive impact 2D materials will have on electronics, especially when linked to optics, is another interest of Kostya Novoselov. He explained that the telecommunications industry is currently trying to move signal processing from electronic signal processing to optical systems. In an interview with IT infrastructure experts BroadGroup [22], Kostya said: "This is because all the interior communications these days are based on fibre optics. And then if we need to do processing, we have to convert it back to electronics, do the processing, convert it back to light, and send it again. The more processing you can do on the optical side, the faster your telecommunication will be. Graphene allows very fast signal processing with the use of silicon photonics. And it has been demonstrated recently by Ericsson. It will be used, and I can't see any other material being as successful for that application as graphene."

In recent years, Kostya has also been working with Manchester-based Nanoco Group PLC, a world leader in the development and manufacture of cadmium-free quantum dots and other types of nano-materials. The company was spun out of the work pioneered by Dr Nigel Pickett and the late Professor Paul O'Brien, at The University of Manchester in 2004 [23]. Quantum dots are about 10 to 100 atoms in diameter and the equivalent to about 1/1000th the width of a human hair. When one of these tiny particles is stimulated by light it can absorb the energy and re-emits the light in a different colour depending on the size of the particle. As a result, quantum dots have offered a new platform technology for LCD displays, lighting and also biomedical applications. In 2018 Nanoco launched a subsidiary, Nanoco 2D Materials, to help pioneer development work in the field of 2D nanoparticles and enable commercial scale-up—and if successful the potential looks exciting.

11.11   Evolution, Not Revolution

Andre Geim believes graphene has gone from lab-to-market in a relative quick time but expectations need to be manged. As an example of a high-tech application, he points to graphene-related technology being used within LSD screens to help reduce temperatures of the screens and the chips by about one degree. Andre admits this adoption is more fast paced evolution than overnight revolution because of the realities of working in the commercial world. However, through constant R&D experimentation—i.e. a "fail fast, learn fast" process—industry is learning to make graphene and associated materials work for them and each success creates a virtuous circle of innovation. "You hear good positive news, this good positive news translates into funding for applied research, and sometimes funding fundamental research, and this spiral continues to propagate. So eventually, there's no doubt about this, we will see principally new products based on 2D materials. I think it's practically inevitable that a certain moment we will see LSD screens and transparent, flexible electronics based on a two-dimensional material."

This expectation matches the analysis of Modor Intelligence (April 2020) [24] who show currently that electronics and telecommunications sectors globally are by far the largest industrial end-users for graphene and its derivative products. Graphene is used or expected to be used in a wide range of applications, including unbreakable touchscreens, transistors, supercharged batteries (i.e. enhanced lithium-ion batteries), optical electronics, printed electronics, and conductive links. The next big end-users, according to Modor, are the defence/aerospace sector, followed by energy, biomedical/healthcare and other niche sectors, such as water filtration.

11.12 Graphene Goes Green: Responding to Climate Change

As society and industry look to transform to a zero-carbon world, graphene and the Graphenes are increasingly being looked at as a way to transform existing technologies and processes. In turn this is providing one of the biggest catalysts to new research and applications.

11.13 Reinventing the Battery

One of the most obvious candidates is energy storage and improving battery technology, too long an innovation laggard. So far, explain Kong et al. (*Nature Nanotechnology*, 2019) [25], graphene has mostly been used as a conductive agent in electrodes to improve rate capability and cyclability because of the material's thermal and electrical conductivity. However, commercial batteries with graphene and graphene–silicon composite electrodes are under development and some battery manufacturers have been leading on graphene battery technology that boast superfast charging speeds when compared to standard lithium-ion batteries and this technology is expected to be adopted in smart phones and even electric-powered vehicles. In 2017, a team of researchers at the Samsung Advanced

Institute of Technology (SAIT) developed so-called "graphene balls", a unique popcorn-shaped battery material that the company claims can offer a 45 per cent increase in capacity and five times faster charging speeds than standard lithium-ion batteries. In an official statement, Samsung said: "… the breakthrough provides promise for the next generation secondary battery market, particularly related to mobile devices and electric vehicles" [26].

While in Manchester, academics are working with an international industrial partner to deliver a breakthrough material to propel the development of supercapacitors. Dr Craig Dawson from the Graphene Engineering Innovation Centre (GEIC) explains: "We fully realise that a simple 'swap in/out of materials' will not offer the performance increases expected; and we also realise that not all 2D materials were created equally—however, by combining our knowledge of these new ultrathin materials with our formulation and engineering experience we are confident of replicating many of the performance gains that have been heralded by the academic research. Put simply, graphene integration into batteries and related devices is not a plug and play. More industry-focused development is required to achieve the breakthroughs needed to deliver the long-anticipated revolution in energy storage technology" [27].

An exemplar of this academic-industry partnership is First Graphene Ltd—an Australian-based business that is one of the GEIC's commercial partners which has signed an exclusive worldwide licensing agreement with The University of Manchester to develop a graphene-hybrid material for use in supercapacitors. These powerful new generation energy storage devices can be used in applications ranging from electric vehicles to elevators and cranes. The licencing agreement is for patented technology for the manufacture of metal oxide decorated graphene materials, using a proprietary electrochemical process. This pioneering application follows published research by two Manchester professors, Robert Dryfe and Ian Kinloch, which revealed how high capacity, microporous materials can be manufactured by the electrochemical processing of graphite raw materials. This process uses transition metal ions to create metal oxide decorated graphene materials, and these have an extremely high gravimetric capacitance, to 500 Farads/g. The breakthrough has led Robert Dryfe to secure support from UK science funding body the Engineering and Physical Sciences Council (EPSRC) for further optimisation of metal oxide-graphene materials. Once this critical phase of research is completed, First Graphene is planning to build a pilot-scale production unit at its laboratories within the GEIC. What is really exciting about this partnership, and in line with the UK government's national industrial strategy, is the anticipation that this will be the first step in volume production in the UK to enable the introduction of these transformative materials to supercapacitor device manufacturers [28].

11.14 Building the Future: Graphene in Construction

According to Chatham House, the international affairs institute, the global production of cement—the "glue" that holds concrete together—accounts for a staggering 8% of the world's CO_2 production. Interestingly, recent experiments with graphene enhanced concrete have been really promising in showing the potential to reduce production levels while meeting demand. Adrian Nixon, editor of the *Nixene Journal*, an independent publication dedicated to graphene and 2D materials science news, has conducted a review of the various studies on adding tiny amounts of graphene and graphene oxide to concrete. Adrian said the addition of just 0.03% graphene powder increased the strength of concrete by a conservative average of 25%. So, bearing in mind worldwide cement production equates to 8% of all global CO_2 emissions, Adrian therefore argues that by improving concrete performance by a quarter through the addition of graphene, we could in turn see this run through the supply chain and potentially deliver a 2% reduction in CO_2 levels [29]. That is an exciting proposition and one that could be debated at great length—but the essential point is this; adding a modest amount of graphene to a building material such as concrete could have a transformational influence on reducing our global carbon levels.

Among those testing this possibility is GEIC partner First Graphene who has worked with the University of Adelaide. Initial test results in 2018 revealed that the addition of 0.03% standard graphene is the optimal quantity of graphene to achieve a 22 to 23

per cent increase in compressive and tensile strength, respectively [30] —in line with Adrian's average. More recent research published in *Construction and Building Materials* (February 2020) showed even more impressive outputs with 0.07% being an optimal dosage for pristine graphene with 34.3 percent and 26.9 pristine graphene enhancement of compressive and tensile strengths at
at 28 days [31]. The work on graphene-enhanced concrete has been further accelerated by GEIC partner, Nationwide Engineering. The construction company is one of the beneficiaries of the European-funded ERDF Bridging the Gap programme at the GEIC and initial trials at lab scale showed significant improvement in compressive and tensile strength with small amounts of graphene oxide added to a concrete recipe pioneered by the firm. This work then led to Innovate UK funding which was shared between Nationwide Engineering, GEIC, and The University of Manchester's Department of Mechanical, Aerospace and Civil Engineering, to deliver large-scale pilot trials. From this work Concretene was born – a new, graphene-enhanced concrete engineered for sustainability. In May 2021, the development team scaled-up from the pilot trials and successfully laid a 700m2 floor slab of a new gym in Amesbury, Wiltshire, with Concretene. Using all conventional equipment and labour, the team used 30% less material for comparable performance to a standard RC30 concrete and removed all steel reinforcement, further reducing associated costs and emissions. In September 2021, Nationwide Engineering poured a concrete slab on an access road adjacent to the GEIC to create a 'living lab' to further test and analyse their product in real world conditions.

How does graphene make a difference? Research has revealed that graphene or related additives like carbon nanotubes have the ability to halt crack propagation in concrete by controlling the nano-sized cracks before they form micro-sized cracks and so greatly improve peak toughness [30]. Interestingly, the GEIC is pursuing a similar infrastructure challenge with Tier 2 partner National Highways to develop a graphene-based bitumen. This new road material would need to be more durable than existing products but also be elastic, so as to survive hot or cold expansion conditions which inevitably lead to surface cracks and ultimately potholes (see Partnership Case Study in Chapter 8).

11.15   Responding to Covid-19

Perhaps one of the most galvanising drivers on graphene research and application since its original isolation has been focusing its capabilities on the response to the coronavirus. "It is crucially important to develop the vaccine against the virus—but, in the meanwhile, we need to apply all the possible measures to minimise the spread of the virus," said Kostya Novoselov in the winter of 2020, who added: "Our task is to make those 'graphene helpers' as efficient as possible. I don't think graphene is a panacea, but it holds a number of attractive characteristics and it definitely will be used—and, in fact, already is being used—to fight the virus, either in diagnostics or in protective gear or as an active ingredient."

Covid-19 is a making a global impact and so we are seeing a global response to this crisis. Massimiliano Papi, Associate Professor at Università Cattolica del Sacro Cuore, and Dr Valentina Palmieri, an expert in nanomaterial synthesis and characterisation and nanoparticle research, have summarised the current state of knowledge on possible applications of graphene to fight Covid-19 (August 2020) [32]. They say: "[The]World Health Organization continues to highlight the need for the prioritization of personal protective equipment supplies for frontline healthcare workers and graphene could be used as coatings of facemasks to minimize the risk of transmission. Graphene textile applications for epidemiological exposure detection and for filtering are possible allies of health systems against pandemic spreading."

While the Graphene Flagship—the EU-funded European-wide research network focusing and progressing graphene research—announced that a team of its experts have united to tackle the effects of present and future pandemics with technologies based on graphene and related materials. They have established the Graphene Flagship Coronavirus Working Group with the aim to establish new connections between researchers, to identify relevant topics for future funding calls, and initiate discussions with funders and stakeholders—the group's ultimate aim is to mobilise use of graphene and related materials in fields such as virology, biosensing and other areas. In its official statement, the organisation said: "The current Covid-19 pandemic brought to light an urgent need to devise new technologies to protect the human body from its immediate environment. Graphene and related materials are promising candidates for the design of a novel generation of surfaces to help deal with the daily challenges posed by Covid-19, as well as similar future diseases" [33].

Also, brand new agencies have been established in the wake of Covid-19. For example, the Advanced Materials Pandemic Taskforce (AMPT) has been set up as a non-political, international group to help combat the Covid-19 pandemic and future public health emergencies at a global level [34]. This new organisation is led by an international steering committee comprising of key leaders in the advanced materials community, including James Baker, co-author of this book. This group also includes representatives from the National University of Singapore, Chalmers Industriteknik, in Sweden, Northwestern University in the US, and the Institute of Photonic Sciences (ICFO) in Spain. The AMPT steering committee says it aims to provide the "vision, strategy, coordination and expert leadership" needed to evaluate and fast-track the use of graphene and other advanced materials to help prevent and manage Covid-19 and future pandemics. At the same time, it will look to develop solutions for the new wave of technological and societal needs that will arise in the aftermath of the coronavirus crisis.

Having graphene being recognised by such global and strategic leadership is helpful to the reputation of advanced materials and also for galvanising research and applications at ground level. For example, graphene is already being used in masks and face coverings. Advanced materials group Versarien have developed a multi-layer mask. And one of the mask's layers features an advanced graphene material—branded Polygrene™—which is said to give robust strength while at the same time providing thermal cooling properties. Using existing production methods, the Polygrene™ layer is blended with a sustainably sourced cellulose-based viscose material mix [35].

While UK-based 2D materials specialists planarTECH has announced an extension of its existing agreement with Thailand-based IDEATI to now include the marketing and distribution of its version of a graphene-enhanced antibacterial face mask. Ted P Thirapatana, Director of IDEATI, is quoted in *Graphene-info* in April 2020 as saying: "This product has been under development for some time, and we sincerely hope it can have a beneficial impact in this time of global crisis. This is our second product launch and we sincerely believe that graphene-enhanced products have reached a tipping point in the market" [36].

The use of fabrics to help safeguard us from Covid-19 is a research is seen a very important. An international team of researchers, including Kostya Novoselov, have reviewed the most advanced antimicrobial finishes on fabrics to protect the wearer against contamination, including the use of graphene. The survey also examined the role of graphene in adoption of wearable textiles. In conclusion to their paper, Nazmul Karim et al. (2020) [37] discuss how Graphenes such as graphene, graphene oxide (GO), reduced GO (rGO), and graphene quantum dots (GQDs) "have shown promise as a new class of broad spectrum antimicrobial agents". Furthermore, Karim et al. also highlight graphene's role in smart wearables as a valuable mechanism to combat Covid-19 and explain: "Recently, we also reported washable, durable, and flexible graphene-based wearable e-textiles, which are highly scalable, cost-effective, and potentially more environmentally friendly than existing metals-based technologies. In addition, graphene and other 2D materials have drawn significant interest in flexible and wearable electronics applications, due to their outstanding electrical, mechanical, and other performance properties. Such properties could also be exploited in heterostructures, where different 2D materials are inkjet printed on top of each other *via* rapid, precise, and reproducible deposition of controlled quantities of 2D materials in a nonimpact, additive patterning, and mask-less approach. Therefore, integrated graphene-based wearable e-textiles for protective clothing could potentiality address current challenges associated with the early detection of highly infectious diseases, ensuring good health and well-being of frontline workers."

Independent of this study, Professor Henry Yi Li, Chair of Textile Science and Engineering at The University of Manchester, has also advocated the role of graphene-enhanced smart materials in safeguarding personal from the coronavirus. Professor Yi Li, who specialises in designing next generation face masks, PPE and smart textiles, says wearable technology has a vital role to play in the "detection, testing and diagnosis" of Covid-19. He explains [38] that by combining e-textile antennas and fabric sensors that feature graphene or other 2D materials it would be possible to create a wearable device to monitor different types of physiological health indicators, such as breathing rates, heart rates, body temperature and sweating rates. This biodata could then be assessed and results feedback to an individual using an appropriate mobile phone app linked to a cloud-based digital health service. This type of technology could be helpful to a person in quarantine awaiting the green light to re-join society—and also, if permission were given, the data could be anonymously shared more widely to provide public health observers real time information on the spread or retreat of a pandemic and help them better manage the public health response. Henry Yi Li adds: "Smart materials and wearable technologies could potentially be plugged into wider health and bio-security systems as part of digitally connected 'smart cities' to help safeguard communities of the future from pandemics like Covid-19."

Graphene is also playing its role in developing biosensor and bioreceptor technologies for testing. For example, an interdisciplinary team of researchers from The University of Manchester has developed a new graphene-based testing system for disease-related antibodies. Initially targeting a kidney disease called membranous nephropathy the new instrument—based on the principle of a quartz-crystal microbalance (QCM) combined with a graphene-based bio-interface—offers a cheap, fast, simple and sensitive alternative to currently available antibody tests. Manchester researcher and entrepreneur Professor Aravind Vijayaraghavan—who also worked on developing the inov-8 graphene-enhanced running shoe (see the Innovation Case Study in Chapter 3)—was the lead investigator on this project and said the system could potentially be adapted to test for antibodies to foreign proteins or viral infections, such as Covid-19. Aravind explained: "Due to the cheap, fast, simple and sensitive nature of our test, we believe that it is ideal for large-scale deployment in response to pandemics in the UK and elsewhere. In particular, the system would offer significant boost to testing capacity in low- and middle income countries and remote locations" [39].

In their summary of graphene's role in combating Covid-19, Professor Papi and Dr Palmieri [32] highlight the high throughput of diagnostics and drug screening based on graphene sensors which "… have been successfully demonstrated and some of these sensors have begun to make their way to the market place." Indeed, researchers in Germany have recently been developing an electronic sensor based on graphene oxide and they claim it can detect bacterial and viral infections such as Covid-19 antibodies in just 15 minutes. The team at the Fraunhofer Institute for Reliability and Microintegration (IZM) in Berlin have been working on the Graph-POC project pre-Covid-19—i.e. since April 2018—using a graphene oxide-based sensor platform. This technology uses a single drop of blood or saliva for an analysis in 15 minutes using a 3D structure of graphene flakes rather than the 2D monolayers used in other sensors [40]. This 3D structure increases the measuring surface and the accuracy of measurements. While US-based Grolltex has developed a graphene-based virus testing platform to help combat Covid-19. The project involves using hand-held reader units and disposable plastic testing chips designed for points of entry including hospitals and venues providing care services [41]. Graphene is also offering huge potential in other areas of health not related to Covid-19, especially in the field of nanomedicine. For example, in March 2021, the spin-out company INBRAIN Neuroelectronics hit the headlines after landing a funding boost for its disruptive graphene-enhanced technology proposition which is intended to overcome the current limitations of metal-based neural interfaces. It is expected that the new graphene devices can help better treat brain disorders such as epilepsy and Parkinson's Disease. The project, involving Kostas Kostarelos, Professor of Nanomedicine at Manchester, received a total of £12 million in funding from a mix of Spanish and German backers—and is one of the largest investments to date in the European medical nanotechnology industry.

In separate research publicised in April 2021, Kostas was part of another European team that demonstrated the potential for graphene to regulate the neurological processes in animals that provoke stressful memories, paving the way for novel therapies for long-term anxiety conditions such as post-traumatic stress disorder (PTSD).

In his famous talk entitled *There's Plenty of Room at the Bottom* [42], the American theoretical physicist Richard Feynman predicted back in 1959 a promising future of nanomaterials. He said: "I can't see exactly what would happen, but I can hardly doubt that when we have some control of the arrangement of things on a small scale we will get an enormously greater range of possible properties that substances can have, and of different things that we can do." Basically, he foresaw the shrinking down of electrical components from the macro scale of his own time to a Wonderland of the future featuring micro- and nanosized machines. The isolation of graphene in 2004 and the subsequent breakthroughs are now taking us to the brink of delivering these exciting and transformative electronic devices and other technologies set to be tranformed by 2D materials. As noted earlier in this chapter, graphene pioneer Andre Geim says this has been a fast-paced evolution and not an overnight revolution—but the speed of lab-to-market delivery has still been impressive for such a young material.

At the macro level, global challenges such as climate change and the Covid-19 pandemic are acting as catalysts to innovation and therefore providing powerful incentives to accelerate the translation of groundbreaking science into new products. The call to "build back better" and the need for sustainable manufacture are definitely factors in helping graphene and the Graphenes to reach a tipping point in delivering next-generation products.

References

1. A. Nixon, Editor of the *Nixene Journal*, interviews one of the busiest researchers in the world, Professor Sir Andre Geim. https://www.nixenepublishing.com/article/adrian-nixon-interviews/.
2. A. Plume. Graphene: Ten years of the 'gold rush'. *Research Trends*, September 2014. https://www.researchtrends.com/issue-38-september-2014/graphene-ten-years-of-the-gold-rush/.
3. Graphene: The transformative new material driving economic growth, Russell Group case study. https://www.russellgroup.ac.uk/policy/case-studies/graphene-the-transformative-new-material-driving-economic-growth/.
3a. The impact of the production and characterisation of graphene, case study submission by The University of Manchester to the Research Exercise Framework (REF) 2014. https://impact.ref.ac.uk/casestudies/CaseStudy.aspx?Id=28174.

4. Electrons flowing like liquid in graphene start a new wave of physics, press release from The University of Manchester, January 2018. https://www.manchester.ac.uk/discover/news/electrons-flowing-like-liquid-in-graphene-start-a-new-wave-of-physics/.
5. Jarillo-Herrero Group, Quantum Nanoelectronics@MIT. http://jarilloherrero.mit.edu/research/.
6. D. H. Freedman. With a simple twist, a 'magic' material is now the big thing in physics, *Quanta Magazine* https://www.quantamagazine.org/how-twisted-graphene-became-the-big-thing-in-physics-20190430/.
7. Manchester-led research offers advance in superconductors with 'twist' in rhombohedral graphite, blog published by Graphene@Manchester, August 2020.
8. J. Barrier, P. Kumaravadivel, R. K. Kumar, L. A. Ponomarenko, N. Xin, M. Holwill, et al. Long-range ballistic transport of Brown-Zak fermions in graphene superlattices, *Nature Communications*, 2020, 11, 5756.
9. J. Baker. Manufacturing sector needs to be 'Graphenes'-ready, *New Statesman*, September 2019. https://www.newstatesman.com/spotlight/manufacturing/2019/09/manufacturing-sector-needs-be-graphenes-ready.
10. M. Jacoby. 2-D materials go beyond graphene, *Chemical & Engineering News* (C&EN), 2017, 95(22). https://cen.acs.org/articles/95/i22/2-D-materials-beyond-graphene.html.
11. V. P. Ningrum, et al. Recent advances in two-dimensional magnets: Physics and devices towards spintronic applications, *Research*, Article no. 1768918. https://spj.sciencemag.org/journals/research/2020/1768918/.
12. M. Gibertini, M. Koperski, A. F. Morpurgo, K. S. Novoselov. Magnetic 2D materials and heterostructures, *Nature Nanotechnology*, 2019, 14(5):408–419.
13. A. L. Sharpe, E. J. Fox, A. W. Barnard, J. Finney, K. Watanabe, T. Taniguchi, M. A. Kastner, D. Goldhaber-Gordon. Emergent ferromagnetism near three-quarters filling in twisted bilayer graphene, *Science,* 2019, 365(6453), 605–608.
14. F. Mouhat, et al. Structure and chemistry of graphene oxide in liquid water from first principles, *Nature Communications*, 2020, 11, Article number 1566, https://www.nature.com/articles/s41467-020-15381-y.
15. Graphene sieve turns seawater into drinking water, press release from The University of Manchester, April 2017, https://www.manchester.

ac.uk/discover/news/graphene-sieve-turns-seawater-into-drinking-water/.

16. X. Huang, et al. Graphene oxide dielectric permittivity at GHz and its applications for wireless humidity sensing, *Scientific Reports*, 2018, 8, Article no. 43. https://www.nature.com/articles/s41598-017-16886-1.

17. Manchester scientists develop graphene sensors that could revolutionise the Internet of Things, press release from The University of Manchester, January 2018. https://www.manchester.ac.uk/discover/news/manchester-scientists-develop-graphene-sensors-that-could-revolutionise-the-internet-of-things/.

18. K. Elibol, et al. Graphene pizza: It is all about the toppings, *Laboratory News*, July 2020. https://www.labnews.co.uk/article/2030752/graphene-pizza-it-is-all-about-the-toppings (taken from "Process pathway controlled evolution of phase and Van-der-Waals epitaxy in in/In_2O_3 on graphene heterostructures", June 2020. https://onlinelibrary.wiley.com/doi/full/10.1002/adfm.202003300).

19. Scientists create fully electronic 2-D spin transistors, *PhyOrg*, September 2019. https://phys.org/news/2019-09-scientists-fully-electronic-d-transistors.html; and *Science Daily*, September 2019. https://www.sciencedaily.com/releases/2019/09/190917115441.htm (and based on *Nano Letters* 2019, 19, 9, 5959–5966. https://doi.org/10.1021/acs.nanolett.9b01611).

20. Graphene and 2D materials could move electronics beyond 'Moore's Law', *Science Daily*, June 2020. https://www.sciencedaily.com/releases/2020/06/200603122949.htm (taken from a press release published by The University of Manchester on 3 June 2020. https://www.manchester.ac.uk/discover/news/graphene-and-2d-materials-could-move-electronics-beyond-moores-law/).

21. Page 27 of the Graphene Flagship annual review for 2019. https://graphene-flagship.eu/graphene/news/the-graphene-flagship-publishes-its-2019-annual-report/.

22. Silicon's final days? An exclusive chat with Nobel Prize winner Sir Konstantin Novoselov, BroadGroup (updated July 2020). https://www.broad-group.com/data/news/documents/b1m08w8xmtvzfg.

23. *Break/though*, 2016, an e-book published by The University of Manchester. http://documents.manchester.ac.uk/display.aspx?DocID=35181.

24. Graphene Market—Growth, Trends and Forecast (2020–2025), Mordor Intelligence, April 2020. https://www.mordorintelligence.com/industry-reports/graphene-market.
25. W. Kong, H. Kum, S. Bae, et al. Path towards graphene commercialization from lab to market, *Nature Nanotechnology,* 2019, **14**, 927–938. https://doi.org/10.1038/s41565-019-0555-2.
26. Samsung develops battery material with 5x faster charging speed, press release published by the Samsung newsroom, November 2017. https://news.samsung.com/global/samsung-develops-battery-material-with-5x-faster-charging-speed.
27. C. Dawson. How graphene innovation could give the UK economy a positive charge, a case study authored in *Spotlight: Energy and Climate Change: Sustainable Politics*, *New Statesman*, 29 November 2019, page 29. https://www.newstatesman.com/sites/default/files/ns_energy_spotlight_supplement_dec_2019.pdf.
28. First Graphene and University to work together to help develop a new graphene-based energy storage material, press release published by The University of Manchester, 23 September 2019. https://www.manchester.ac.uk/discover/news/first-graphene-and-university-to-work-together-to-help-develop-a-new-graphene-based-energy-storage-material/.
29. D. Nelson. Graphene in concrete, Adrian Nixon, editor of the *Nixene Journal* interviewed, contributing editor and project manager at the *Nixene Journal*. https://www.youtube.com/watch?v=0vraTJrqErk&t=168s.
30. T. Barkan. Concrete graphene applications, literally!, 10 January 2018, *The Graphene Council* web site. https://www.thegraphenecouncil.org/blogpost/1501180/292599/Concrete-Graphene-Applications-Literally.
31. V. D. Ho, et al. Electrochemically produced graphene with ultra large particles enhances mechanical properties of Portland cement mortar, *Construction and Building Materials,* 20 February 2020, 234, 117403. https://www.sciencedirect.com/science/article/abs/pii/S0950061819328557?via%3Dihub.
32. V. Palmier, M. Papiab. Can graphene take part in the fight against COVID-19?' *Nano Today*, August 2020, 33. https://www.sciencedirect.com/science/article/pii/S1748013220300529; https://doi.org/10.1016/j.nantod.2020.100883.

33. Graphene Flagship launches COVID-19 task force, press statement published by Graphene Flagship, June 2020. http://graphene-flagship.eu/news/Pages/COVID-Taskforce-Announcement.aspx.
34. Official web site for Advanced Materials Pandemic Taskforce (AMPT). https://www.amptnetwork.com/.
35. Versarien corporate web site promoting its graphene face mask. http://www.versarien.com/mask.
36. Planartech and IDEATI launch graphene-enhanced antibacterial face masks, *Graphene-info*, April 2020. https://www.graphene-info.com/planartech-and-ideati-launch-graphene-enhanced-antibacterial-face-masks.
37. N. Karim, S. Afroj, K. Lloyd, L. C. Oaten, D. V. Andreeva, C. Carr, et al. Sustainable personal protective clothing for healthcare applications: a review, *CS Nano*, 31 August 2020. https://doi.org/10.1021/acsnano.0c05537.
38. Taken from an (unpublished) "flash lecture" given by Professor Henry Yi Li as part of the Catalyst Catalyst campaign run by The University of Manchester, September 2020.
39. New graphene-based antibody test developed for detecting kidney disease, press release from The University of Manchester, 26 October 2020. https://www.manchester.ac.uk/discover/news/new-graphene-based-antibody-test-developed-for-detecting-kidney-disease/.
40. N. Flaherty. Graphene sensor detects Covid 19 antibodies in minutes, *eeNews* (Europe), 3 August 2020. https://www.eenewseurope.com/news/graphene-sensor-detects-covid-19-antibodies-minutes.
41. S. Moore. How graphene sensors from Grolltex are being used in the fight against COVID-19, *AZO Nano*, 23 April 2020. https://www.azonano.com/news.aspx?newsID=3725637.
42. R. Feynman. There's plenty of room at the bottom, annual meeting of American Physical Society, California Institute of Technology (1959). *Google scholar*_https://scholar.google.com/scholar?q=Feynman,%20R.,%20Theres%20Plenty%20of%20Room%20at%20the%20Bottom,%20Annual%20meeting%20of%20American%20Physical%20Society,%20California%20Institute%20of%20Technology%20.

Chapter 12

Future Histories: Graphene Innovation after Covid-19
By James Baker

As I write this final chapter, we are heading towards the end of 2021 and unfortunately experiencing the on-going reverberations of the Covid-19 pandemic. We had feared when this started, many months previously, the coronavirus crisis would have a huge impact and slowdown of the potential development of graphene products and applications.

However, what we have seen over this period is arguably unprecedented, in terms of an increased pace and acceleration of innovation around graphene and 2D materials applications. We have actually grown the number of industrial partners that we are engaging with through the GEIC and many projects are underway following the return to the labs in mid-2020. We are starting to see, from the early collaborative projects being delivered with these partners, the first products starting to hit the marketplace, including a number that we have discussed as case studies throughout this book.

Not only are we seeing an increase in graphene products and applications already this year, but we are seeing other Graphenes being developed again with significant market opportunities. These includes boron nitride (BN), molybdenum di-sulphide (MoS_2) and MXenes to name just a few. I am confident that you will start to see

---

Graphene: The Route to Commercialisation
James Baker and James Tallentire
Copyright © 2022 Jenny Stanford Publishing Pte. Ltd.
ISBN 978-981-4877-87-9 (Hardcover), 978-1-003-20027-7 (eBook)
www.jennystanford.com

an increasing number of these Graphenes as we progress into the future. Graphene itself also continues at a pace, not only in the UK but increasingly you can see examples of graphene products and applications hitting the marketplace.

Another key phenomenon we are now starting to see compared to the early days of graphene—and could arguably be described as a progressive move along the Gartner hype cycle—are new products appearing on the marketplace that contain graphene but without the marketing or description to identify its presence. People are therefore starting to use graphene because of the performance benefit not just because of the marketing advantage. While there are still companies using the marketing benefits of being associated with graphene, to me, this is more evidence that shows that graphene is starting to become part of mainstream products and applications in the future.

When we decided to write this book, a key driver was that of defining "future histories" for graphene and the pioneering Manchester Model of Innovation that we have built to commercialise 2D materials and has been inspired by my earlier career experiences. When we look back and reflect on this book, I believe we will see that the last two to three years have been a significant inflection point in the development of graphene and 2D materials products and applications. In what is probably unprecedented timescales, when you consider for example carbon-fibre took over 25 years for the first products and applications to hit the marketplace. We are already seeing with graphene that despite being "17 years young" there is a real growth in products and applications in the marketplace. While many of these are still quite niche applications there are a number of examples where graphene has played a significant part in a large number of mainstream products, from mobile phones to car components to paints and formulations.

12.1 Demand-Led Approach

What we have built into our Manchester model is the capability to respond to demand and be sure to meet the needs of external stakeholders and the forces they need to respond to. This demand and output relationship is shown in the graphic below (Fig. 12.1) which highlights the Graphene@Manchester research and

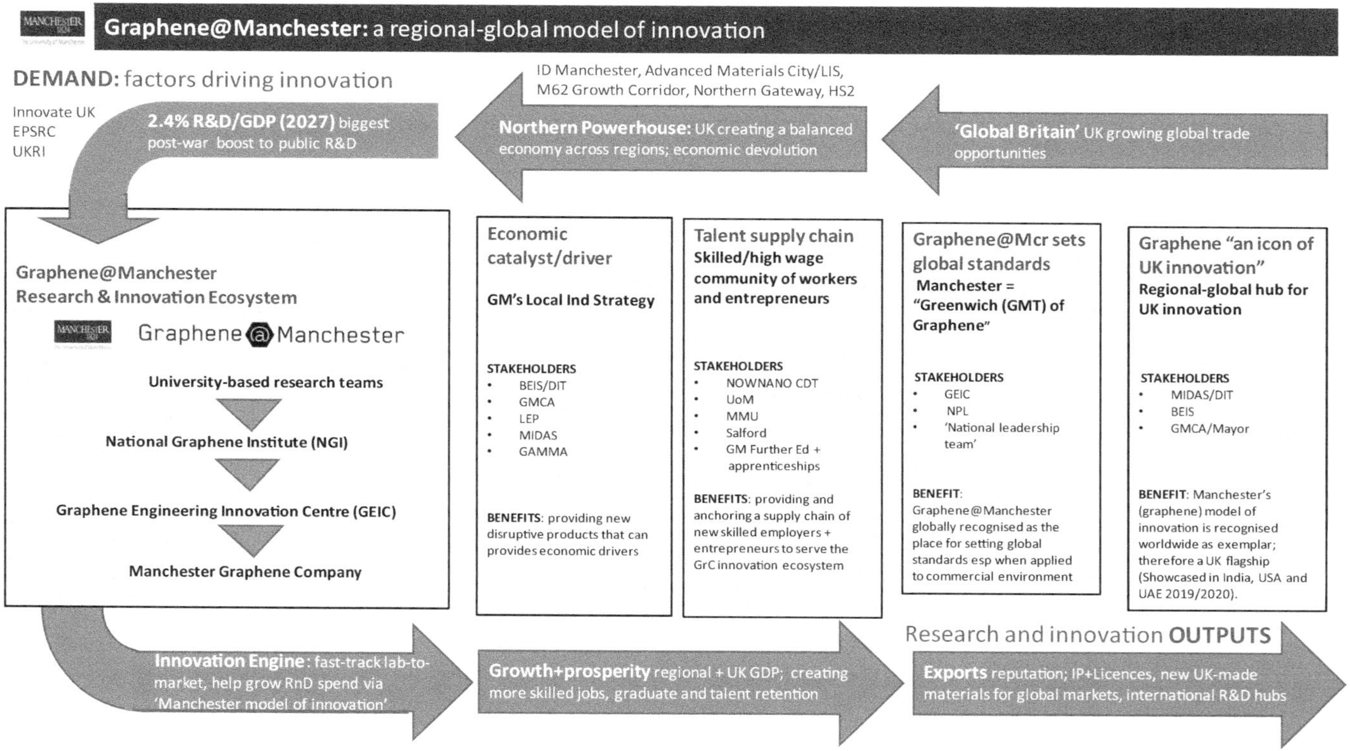

Demand-Led Approach | 165

Figure 12.1 Graphene@Manchester: Manchester's regional-global model of innovation which shows the push (outputs) and pull (demand) relationships. From James Tallentire/James Baker © 2020.

innovation community (including the NGI, GEIC and the Manchester Graphene Company) as an "innovation engine" and its relationship with the wider innovation ecosystem, including expectations to meet the needs of the regional economy, such as economic growth and supporting a talent supply chain; as well as global outputs including international standards for graphene (working with the National Physical Laboratory) and showcasing graphene as an "icon of UK innovation" to key markets like North America or Asia. The factors driving demand will evolve and change—who was expecting the impact of a global pandemic at the start of 2020 for example—but we are confident we can be highly reactive when we need to be but can also provide leadership in bringing transformational advanced materials to the marketplace.

12.2 At the Tipping Point

As we move from 2021 into 2022 and hopefully start to emerge from the global Covid-19 pandemic, I really believe that we will see graphene achieve the "tipping point" of commercialisation and that we will see an increasing and ever-growing number of products and applications where graphene plays a key role in the future. Interestingly English-born Canadian journalist and author Malcolm Gladwell compared the phenomenon of a "tipping point" to an epidemic that once it catches hold, for whatever reason, spreads like a virus. In his book *The Tipping Point: How Little Things Can Make a Big Difference* [1], Malcolm Gladwell, defined the idea of a "tipping point" as "the moment of critical mass, the threshold, the boiling point".

In addition to any post Covid-19 recovery, another factor I believe will take us to the moment of critical mass for widespread adoption of graphene and the Graphenes will be the response to the sustainability agenda and the drive to achieve carbon neutrality across the world economy. Graphene can play a critical role in support of this ambitious agenda, for example, it can be added to materials to improve their recyclability (e.g. old tyres or plastics) through to graphene being used for its own unique environmental benefits (lightweighting and multifunctional capabilities) to deliver transformational and disruptive performance advantages in a new range of products.

12.3 Future Histories

It has been a very exciting period to be associated with graphene and as we look to the challenges ahead there is real confidence that the tipping point for mainstream commercialisation is now within reach. However, to avoid hubris we should perhaps reflect on the lessons we can learn from the recent past. A good example is the Manchester computer pioneers that were discussed in Chapter 10 and how their technological legacy for the city, the region and the UK as a whole was eventually lost to the USA and then Japan within a few decades. Today a replica of their landmark contribution is housed in a museum while the global home for high technology and innovation is in Silicon Valley, California. I am also reminded of a blog I authored entitled "Engineering an Olympic legacy", which was commissioned by the *Huffington Post* just after the hugely popular 2012 Olympic Games, successfully hosted in London [2]. In my article, I said the "engineering legacy" based on the fantastic innovation the global competition had inspired would be just as important and almost certainly longer lasting than the sporting legacy. What I call our "future histories" is making sure that what we are doing today resonates with future generations and especially in the place where it was first originated. I sincerely believe that graphene will be a future history for Greater Manchester as the "home of graphene" for a long time to come.

References

1. M. Gladwell. *The Tipping Point: How Little Things Can Make a Big Difference*, 2000, Little, Brown and Company (originally published in hardback). https://www.worldcat.org/title/tipping-point-how-little-things-can-make-a-big-difference/oclc/55586972.

2. J. Baker. Engineering an Olympic legacy, Huffington Post, 20 September 2012. https://www.huffingtonpost.co.uk/james-baker/engineering-an-olympic-le_b_1895930.html.

Index

academic-industry ecosystem 87
Advanced Machinery and Productivity Institute (AMPI) 83
Advanced Materials Pandemic Taskforce (AMPT) 154
Advanced Research Projects Agency (ARPA) 113
Advanced Technology Centre (ATC) at BAE Systems 14, 16–19, 21, 23, 26
aerospace 2–4, 30, 34, 36–38, 41, 46, 50, 80–81, 89, 100, 130–131
agriculture 90, 109
aircraft
  fuel-efficient 35
  next-generation 35
  performance 38
  plastic 35
  technology 38
  wing 133
applications
  biomedical 34, 148
  energy 34
  high-tech 148
  industrial 136, 145
  medical 124
  niche 164
  patentable 62
auto industry 39, 80
automotive industry 91
autonomous navigation 17–18
aviation industry 38

batteries 34, 40, 131, 149–150
  standard lithium-ion 149–150

battery market, next-generation secondary 150
biotechnology 78, 124
  industrial 78
boron nitride (BN) 102, 141–142, 163
Bowler Wildcat 17, 18, 20, 22, 26
Briggs Automotive Company (BAC) 40–41, 127
British Aerospace organisations 14
Brown-Zak fermions 140–141
business challenges 71, 92
businesses, graphene-based 76

carbon 101, 142, 146
carbon fibre 3, 34, 38, 50
  graphene-enhanced 35, 41
carbon films, thin 28
cars 19, 40
cement production 151
chemical vapour deposition (CVD) 99
China 50, 88, 90, 127, 130–131, 133, 136
CO_2 production 151
collaboration 4, 17, 19–20, 25, 39, 43–44, 55, 57, 63, 68, 87, 93, 97
commercialisation 2, 4–5, 13, 16, 26, 45, 51–53, 55, 58, 67, 77–78, 88, 108, 115–117, 120, 136, 166
companies
  biopharmaceutical 90
  government-owned 90
  multi-media 7

startup 79, 109
web 7
computer games 147
computer hardware 80
computer programmers 106
computers
 first programmable 49
 first stored-program electronic 112
 first stored programmed 115
 quantum 138
computing 61, 114, 116
computing technology 113
copper 2, 30, 33
cottage industry 95
Covid-19 41, 92, 105, 125, 153-157, 163

defence industry/market 2, 14, 21, 34
defence sector 14, 19, 21
Defense Advanced Research Projects Agency (DARPA) 22-23, 56, 68, 113
Defense Advanced Research Projects Agency (United States) 68
demand-led approach 164-165
designer materials 30, 102, 143
driving-aids market 17

economic growth 77, 79, 136, 166
electric field effect 28
electricity 146
electronic devices 138, 140-141
electronics 63, 70, 80, 90, 105, 127, 145-147, 149
 wearable 155
electrons 139, 141, 145-146
energy production, sustainable 144
energy storage devices 40, 150

Engineering and Physical Sciences Council (EPSRC) 60-61, 151, 165
European medical nanotechnology industry 157
European Regional Development Fund (ERDF) 61, 69, 94

ferromagnets 143-144
food industry 90

Gartner Innovation Cycle 51
Geim, Andre 26-29, 31, 51, 60, 88, 99, 104, 108, 118-119, 121, 134-135, 140, 146, 148, 158
graphene 1-5, 9-10, 13, 15-16, 20, 25-31, 33-46, 49, 51-52, 55-56, 58-64, 67-68, 70-71, 75, 79-82, 84, 87-88, 90-97, 99-102, 104-105, 108-109, 111-112, 117-124, 127-139, 141-142, 144-149, 151-158, 163-167
Graphene, Advanced Materials and Manufacturing Alliance (GAMMA) 84, 165
graphene
 all-carbon analog 142
 in China and India 130-131, 133
 commercialisation 97
 commercialisation of 3-4, 6, 9, 15-16, 23, 25, 41, 59-60, 64, 67-68, 71, 75, 77-78, 84, 87, 91, 100, 103, 135
 encapsulated 138
 mini 123
 potential applications of 36, 39, 81
 pristine 152
 twisted 138
 twisted bilayer 139-140

Index

graphene accelerator 22
graphene analogs, single-element 101, 142
graphene applications 91–93, 103, 153
graphene balls 150
graphene-based bio-interface 156
graphene-based bitumen 93, 152
graphene-based materials 140, 144
graphene-based virus testing platform 157
graphene-based wearable e-textiles
  flexible 155
  integrated 155
graphene-bodied car 40
graphene devices 157
graphene economy 122
Graphene Engineering Innovation Centre (GEIC) 5, 22, 41, 56–58, 67–72, 75–76, 78, 83, 88–95, 103, 105, 107, 109, 122, 131, 150–152, 163, 165–166
graphene-enhanced antibacterial face mask 154
graphene-enhanced BAC Mono car 89
graphene-enhanced carbon fiber 127
graphene-enhanced rubber 132
graphene-enhanced running shoe 156
graphene-enhanced smart materials 155
graphene entrepreneurs 106
graphene films 101
graphene flakes 157
graphene forms 96
Graphene Hackathon 105–107
graphene humidity sensors 145
graphene-hybrid material 150
graphene ink application 105
graphene-ink products 107
graphene innovation 67, 105, 163
graphene innovation ecosystem 20, 83
graphene innovators 117
graphene layers 99, 138–139, 141
graphene manufacture 88, 96
graphene membranes 25, 145
graphene nanoplatelets 99
graphene nanoplates 101
graphene oxide 99, 101, 144–145, 151–152, 155, 157
  reduced 101
graphene oxide membranes 144
graphene powder 151
graphene products 71, 163–164
graphene quantum dots (GQDs) 155
graphene-related products 97
graphene research 28, 61, 63, 67, 136–139, 153
graphene sensors 109, 156
graphene sheets 28, 33
  defect-free monolayer 30
graphene shoe 46
graphene spintronics 147
graphene straw 25
graphene superlattices 138
graphene supernova 122
graphene supply chain 97
graphene-toned grey paint 62
graphite 26–27, 30, 96, 140
  rhombohedral 139–140

heterostructures 30, 100, 102, 142–143, 146–147, 155
  flexible 145
high value manufacturing (HVM) 56, 71, 83

India 127, 130–131, 133, 165
intellectual property 20, 77

Internet of Things (IoT) 16, 145, 147

Japan 10, 49, 116, 127, 167

Manchester Model of Innovation 6, 54–55, 57–58, 71, 87, 95, 99–100, 110, 127, 164
metal oxide 150
metal oxide-graphene materials 151
mobile phone products 52
MXenes 101, 104, 142, 163

National Graphene Institute (NGI) 5, 44, 55–56, 58–64, 67–68, 70, 72, 75, 79, 83, 88–89, 92, 96, 101, 104, 122, 131, 165–166
National Institute of Standards and Technology (NIST) 132
National Physical Laboratories (NPL) 57, 96, 101, 132, 165
National Research Development Corporation (NRDC) 113
Nobel Prize 28–30, 51, 60, 118–120, 135, 137
Novoselov, Konstantin (Kostya) 26, 28, 31, 51, 55, 60, 62, 65, 88, 105, 119, 131, 134–135, 137, 147, 153, 155, 159–160

open innovation 4, 6, 13–16, 20–21, 23, 25–26
original equipment manufacturers (OEMs) 94

patents 20, 92
post-traumatic stress disorder (PTSD) 157
process innovation 45

quantum dots 148

quartz-crystal microbalance (QCM) 156

sectors
  automotive 19, 40, 127
  civil nuclear 125
  commercial 3, 19
  defence/aerospace 149
  education 3, 6, 26
  energy 130
  marine 90
  medical 34
  telecommunications 149
semiconductor industry 127, 143
sensors 17–18, 92, 143, 156–157
silicon 30, 50, 136, 149
silicon photonics 148
small and medium enterprises (SMEs) 20–21, 55, 80, 94
solar power generation 90
South Korea 127, 136
spin-orbit coupling 146
spintronics 145–147
sports industry 45
start-ups 75–76, 107
  innovative 21
strain sensors, thin graphene 107
supercapacitors 34, 40, 150
supercomputer 113
superconductivity 139–140, 146
superconductors 138–139
superlattices, graphene-boron nitride 140
supply chain 1, 14–16, 20–22, 37, 41, 46, 55, 57, 78, 80, 95, 99, 110, 151, 165
sustainability 34, 43, 94, 152, 166
system readiness level (SRL) 52, 54, 103

technology
  autonomous 17
  battery 149

bioreceptor 156
energy storage 150
graphene-based 46
graphene battery 149
magnetic RAM 147
wireless 145
technology readiness level (TRLs) 52–54, 79, 89, 103
telecommunications industry 147
textiles 80–81, 84, 90
transistors 27, 141, 143, 146, 149
transition metal dichalcogenide (TMD) 102, 142, 146
Turing, Alan 10, 112, 129
twisted bilayer graphene layers, ferromagnetism in 144
Twistronics 138, 139

two dimensional material/s (2D material/s) 2–3, 9, 13, 16, 20, 25–26, 28, 30, 38, 40, 41, 46, 55, 57, 62–63, 68, 71, 84, 87, 89, 90–91, 94, 99, 100–104, 108–110, 117, 121–122, 131, 135–136, 138–145, 147–151, 154–156, 158–160, 163–164

unmanned aerial vehicles (UAVs) 4, 37–38
uranium 139

van der Waals heterostructures 102, 138, 144–146
viral infections 156–157
virus 153, 166